DETECTION OF EXPLOSIVES FOR COMMERCIAL AVIATION SECURITY

Committee on
Commercial Aviation Security

National Materials Advisory Board
Commission on Engineering and Technical Systems
National Research Council

Publication NMAB-471
National Academy Press
1993

NOTICE: The project that is the subject of this report was approved by the Governing Board of the National Research Council, whose members are drawn from the councils of the National Academy of Sciences, the National Academy of Engineering, and the Institute of Medicine. The members of the committee responsible for the report were chosen for their special competences and with regard for appropriate balance.

This report has been reviewed by a group other than the authors according to procedures approved by a Report Review Committee consisting of members of the National Academy of Sciences, the National Academy of Engineering, and the Institute of Medicine.

The National Academy of Sciences is a private, nonprofit, self-perpetuating society of distinguished scholars engaged in scientific and engineering research, dedicated to the furtherance of science and technology and to their use for the general welfare. Upon the authority of the charter granted to it by the Congress in 1863, the Academy has a mandate that requires it to advise the federal government on scientific and technical matters. Dr. Frank Press is president of the National Academy of Sciences.

The National Academy of Engineering was established in 1964, under the charter of the National Academy of Sciences, as a parallel organization of outstanding engineers. It is autonomous in its administration and in the selection of its members, sharing with the National Academy of Sciences the responsibility for advising the federal government. The National Academy of Engineering also sponsors engineering programs aimed at meeting national needs, encourages education and research, and recognizes the superior achievements of engineers. Dr. Robert M. White is president of the National Academy of Engineering.

The Institute of Medicine was established in 1970 by the National Academy of Sciences to secure the services of eminent members of appropriate professions in the examination of policy matters pertaining to the health of the public. The Institute acts under the responsibility given to the National Academy of Sciences by its congressional charter to be an adviser to the federal government and, upon its own initiative, to identify issues of medical care, research, and education. Dr. Kenneth I. Shine is President of the Institute of Medicine.

The National Research Council was organized by the National Academy of Sciences in 1916 to associate the broad community of science and technology with the Academy's purposes of furthering knowledge and advising the federal government. Functioning in accordance with general policies determined by the Academy, the Council has become the principal operating agency of both the National Academy of Sciences and the National Academy of Engineering in providing services to the government, the public, and the scientific and engineering communities. The Council is administered jointly by both Academies and the Institute of Medicine. Dr. Frank Press and Dr. Robert M. White are chairman and vice chairman, respectively, of the National Research Council.

This study by the National Materials Advisory Board was conducted under Contract No. DTFA903-88-C-00021 with the Federal Aviation Administration, U.S. Department of Transportation.

Library of Congress Catalog Number 93-84550
International Standard Book Number 0-309-04945-8

This report is available from:

National Academy Press
2101 Constitution Avenue, NW,
P.O. Box 285
Washington, DC 20055
1-800-624-6242
202-334-3313 (In Washington Metropolitan Area)

B-169

Printed in the United States of America

COMMITTEE ON COMMERCIAL AVIATION SECURITY

Chairman

JOHN D. BALDESCHWIELER, California Institute of Technology, Pasadena

Members

ALFRED BLUMSTEIN, Carnegie Mellon University, Pittsburgh, Pennsylvania

STANLEY S. HANNA, Stanford University, Stanford, California

DAVID MILLIGAN, Abbott Laboratories, Abbott Park, Illinois

NORMAN SLAGG, U.S. Army ARDEC, Picatinny Arsenal, Dover, New Jersey

MICHAEL STORY, Northwest Instrument Systems, Inc., Richland, Washington

HARVEY E. WEGNER, (Retired) Brookhaven National Laboratory, Upton, New York

Liaison Representatives

LEO T. POWELL, Jr., FAA Technical Center, Atlantic City, New Jersey

Technical Consultants

JOSEPH A. NAVARRO, JAN Associates, Inc., Bethesda, Maryland

JEFFERY EBERHARD, GE Corporate R&D, Schenectady, New York

NMAB Staff

ROBERT E. SCHAFRIK, Acting Director

JANICE M. PRISCO, Project Assistant

PANEL ON TEST PROTOCOL & PERFORMANCE CRITERIA

Chairman

MICHAEL STORY, Northwest Instrument Systems, Inc., Richland, Washington

Members

JONATHAN W. AMY, West Lafayette, Indiana

ARTHUR FRIES, Institute for Defense Analyses, Alexandria, Virginia

BRUCE R. KOWALSKI, University of Washington, Seattle

HAROLD MCNAIR, Virginia Polytechnic Institute and State University, Blacksburg

DAVID MILLIGAN, Abbott Laboratory, Abbott, Illinois

HARVEY E. WEGNER, (Retired) Brookhaven National Laboratories, Upton, New York

PANEL ON SYSTEMS ARCHITECTURE

Chairman

JOHN D. BALDESCHWIELER, California Institute of Technology, Pasadena

Members

ALFRED BLUMSTEIN, Carnegie-Mellon University, Pittsburgh, Pennsylvania

STANLEY M. HANNA, Stanford University, Stanford, California

WILFRED A. JACKSON, University of North Dakota, Grand Forks

RICHARD H. JUDY, Richard H. Judy & Associates, Miami, Florida

JOHN R. ORR, Delta Airlines, Atlanta, Georgia

NORMAN SLAGG, U.S. Army ARDEC, Picatinny Arsenal, Dover, New Jersey

ABSTRACT

(U) The threat posed to commercial aviation by small, concealed explosive devices is particularly severe since they are difficult to detect using current techniques and can readily cause tremendous destruction and loss of life. Protecting air travelers from terrorist actions is an essential mission of the Federal Aviation Administration (FAA). The FAA plays a critical role in defining the terrorist threat, stimulating the development of explosive detection devices and systems, and in regulating their use.

(U) The key issues for the FAA regarding explosive detection technology are:

- What can the different detection methods do in principle?
- What can they actually do in practice?
- How can the different methods be best employed to counter the terrorist threat?

(U) This report of the Committee on Commercial Aviation Security addresses the above issues from a detection technology perspective. It discusses and assesses system considerations, testing protocols and performance criteria, and recent explosive detection technology developments.

DEDICATION

This report is dedicated to the memory of Dr. John Sheehan, who was Emeritus Professor of Organic Chemistry at MIT, a Member of the National Academy of Sciences and a former member of this committee. Dr. Sheehan had many accomplishments in his lifetime. He is perhaps best known as being the first to synthesize penicillin. In 1941 he was codeveloper of the large scale method for manufacturing the highly energetic "plastic" explosive, known as RDX. His contributions to this committee, in the aftermath of the tragic bombing of Pan Am 103, transcended his knowledge of the chemistry and synthesis of energetic materials. He worked selfishly and tirelessly in helping to define, for the FAA's Security Technology Program, the "right things to do" to prevent another Pan Am 103. He significantly contributed to this committee's previous report *Reducing the Risk of Explosives on Commercial Aircraft*. Many of the insights provided by Dr. Sheehan have been extended in this report. He set an example of service and dedication for all of us to follow.

John D. Baldeschwieler
Chairman
Committee on Commercial Aviation Security

PREFACE

Since its creation in 1958, the Federal Aviation Administration (FAA) of the Department of Transportation has had the responsibility for ensuring the safety of air travel. The FAA establishes security requirements in response to an assessment of a variety of threats, and then ensures compliance. An important FAA goal is reducing future vulnerability of the civil air transport system to terrorist threats by employing appropriate procedural and technical means to detect and counter the threats.

The development of systems and devices to prevent or deter hijacking or sabotage against civil aviation was first authorized in the Air Transportation Security Act of 1974[1], also known as the "Antihijacking Act of 1974." As part of its obligation under this Act, the FAA began a research and development program that emphasized the development of devices to protect air travelers against acts of criminal violence and aircraft piracy.

The FAA's role in aviation security was expanded in 1985 with the passage of the International Security and Development Cooperation Act[2]. The FAA was required to assess the adequacy of security at foreign airports served by U.S. carriers, and the security procedures of foreign air carriers flying to and from the United States. This law also provided for the expansion of the FAA's research and development program.

In March 1988, the FAA requested that the National Research Council's National Materials Advisory Board (NMAB), of the Commission on Engineering and Technical Systems initiate a study to assist the FAA Technical Center in evaluating near-term and long-term technical programs relating to instrumental methods for detecting concealed explosives, with the emphasis on highly energetic, "plastic" explosive materials in checked baggage. These methods exploit one or more material properties of an explosive for detection purposes. Soon afterward, in December 1988, the critical national importance of explosive detection was highlighted by the terrorist bombing of Pan American Flight 103 over Lockerbie, Scotland.

---

[1] Public Law 93-366; August 5, 1974.

[2] Public Law 99-83; August 8, 1985

The NMAB established the Committee on Commercial Aviation Security, composed of ten experts in the areas of chemistry, physics, materials science, explosive materials, sophisticated analytical instrumentation, forensic science, and ordnance. The committee met eight times, and was briefed about current and proposed programs by FAA officials, FAA-funded contractors, and others. The committee was apprised of the efforts of other federal agencies concerned with related problems. An unclassified, limited-attendance workshop was conducted on "Instrumental Methods for Detection of Explosives" to elicit new ideas from the community at large. This committee published its findings in the 1990 report, *Reducing the Risk of Explosives on Commercial Aircraft(U)*, NMAB-463 (classified Confidential).[3]

The committee's primary conclusion was that "there does not appear to be any single detection technology that can provide levels of sensitivity and specificity that will have both a significant effect on reducing the threat and an acceptable impact on airport operations." The report then analyzed a number of detection technologies and provided broad priorities for allocation of funding.

The committee's recommendations were:

1. ***Define a search strategy to optimize the mix of technologies available***. This recommendation leads to consideration of a systems approach which employs layers of different devices which have orthogonal detection capability.

2. ***Implement low-technology improvements***. These methods include positive bag-to-passenger matching and passenger profiling/interviewing.

3. ***Define performance criteria for detection systems***. The committee suggested, as a guide for an overall systems architecture, a minimum level of sensitivity and specificity, a minimum detectable amount of plastic explosives, vapor detection sensitivity range, and bag through-put rate.

4. ***Explore reinforcing baggage containers to increase the maximum charge that can be contained at altitude***. Straightforward reinforcing of baggage containers, as well as use of compressible padding material, were suggested to harden baggage containers against small explosive charges. Improvements to the aircraft fuselage were suggested as a long term effort.

---

[3]A Summary (unclassified) of the report was also published. It contains the report's Abstract, Preface, and Executive Summary.

5. ***Establish standardized operational tests for all explosives detection systems.*** A FAA standardized test facility was recommended. A three phase testing approach was suggested: determine minimum detection capability of the system using pure explosive materials; test the performance of the system using pure material plus likely interferrents; and conduct field tests under airport operational conditions with a variety of bags containing explosives, interferrents, and no explosives.

6. ***Develop standard positive controls for solid and vapor detectors.*** The reliable generation of very minute, defined quantities of explosive vapor was found to be a difficult problem, but essential to providing a positive control for routine checks of instruments.

7. ***Take advantage of systems integration opportunities for vapor detectors.*** The best stages from several commercial instruments could be combined by an integrating contractor to produce an overall superior instrument.

8. ***Explore the concept of tagging explosives and detonators to make them easily detectable.*** Small quantities of materials could be added to explosives and detonators to make them more observable by relatively inexpensive means.

9. ***Continue to support the exploration and development of new methods that may be applicable to explosives detection.*** The FAA was found to have either supported or monitored the appropriate technologies; continued funding of advanced techniques to provide future options was deemed to be very desirable.

Program priorities to implement the above recommendations included:

- Establish an explosive detection systems analysis and architecture group.
- Demonstrate passenger/luggage correlation schemes.
- Solicit and fund proposals for an aircraft hardening analysis.
- Establish an operational testing facility.
- Solicit and fund proposals for developing positive controls for bulk and vapor phase systems.
- Select a prime contractor or systems architect for vapor phase systems.
- Solicit and fund proposals to demonstrate explosive tagging schemes.
- Solicit and fund exploratory research proposals for new methods of explosive detection.

During the course of the previous NMAB study, in response to the tragic loss of Pan Am 103, Presidential Executive Order 12686 established the "President's Commission on Aviation Security and Terrorism." One of the main recommendations in the R&D area was:[4] "FAA should undertake a vigorous effort to marshal the necessary expertise to develop and test effective explosive-detection systems."

The Aviation Security Improvement Act of 1990[5] was enacted, in large measure, to implement the recommendations of the President's Commission. The Act directs the FAA to[6] "accelerate and expand the research, development, and the implementation of technologies and procedures to counteract terrorist acts against civil aviation." The Act further requires that the FAA not mandate the deployment or purchase of explosive detection equipment unless[7] "... based on the results of tests conducted pursuant to protocols developed in consultation with experts from outside the Federal Aviation Administration, such equipment alone or as part of an integrated system can detect under realistic air carrier operating conditions ... explosive material which would be likely to be used to cause catastrophic damage to commercial aircraft."

The FAA requested that the Committee on Commercial Aviation Security be extended to provide advice regarding the implementation of the Aviation Security Improvement Act of 1990 in two key areas: systems analysis and architecture for explosive detection systems that could inspect passenger baggage, and the development of test protocols and performance criteria for such systems. The committee formed two panels corresponding to these areas. The panels met concurrently. Just as before, the committee and panels were briefed about current and proposed programs by FAA officials, FAA-funded contractors, and others. They were updated regarding the efforts of other federal agencies concerned with related problems. Visits to several facilities were made to gain first-hand knowledge about program details and representatives of the FAA periodically reported testing results. The committee and panels met nine times between November 1990 and September 1992.

During the course of this study, much was learned about the complexity and variability of this environment in which an explosive detection system must successfully operate, as well as the strengths and weaknesses of the

---

[4]Report on the President's Commission on Aviation Security and Terrorism, May 15, 1990, page 122.

[5]Public Law 101-604, November 16, 1990.

[6]Public Law 101-604, Section 107, "Research and Development, Part (3), "Program to Accelerate Research."

[7]Public Law 101-604, Section 108, "Deployment of Explosive Detection Equipment."

different detection approaches. The committee concludes that the effective detection of small quantities of high performance "plastic" explosive materials in a busy commercial airport environment continues to be an extremely difficult problem, most likely requiring more than one detection instrument. Practical issues of affordability, reliability, maintainability, etc., will assume major importance as the technology matures to the point that an informed implementation decision can be made.

A key recommendation addresses the importance of developing powerful simulation tools to allow users to experiment with the architecture of an explosive detection system, and to gain an understanding of the effect that various changes would have on airport operations and system effectiveness. Another key recommendation emphasizes the need for a comprehensive explosive detection system (EDS) certification program that establishes requirements, verifies vendor specifications, certifies operational performance, and monitors systems deployed in the field. The regulatory approach should provide for sustained integrity of explosive detection systems while encouraging technological and operational evolution.

The committee was consulted on a general protocol for testing instrumental methods that detect bulk properties of explosives. The protocol, contained in Appendix A, is a starting point for developing detailed test procedures to certify or verify the performance of a system or device.

The final section of the report provides an update on some specific technologies currently under development for the detection of explosive materials. Broad investment strategy priorities are suggested to assist the FAA in balancing the effort required for the many technology options.

This report is written to be a companion to the previous report. To the extent possible, there is little duplication of information in this report. Where appropriate, the information in the previous report has been updated, particularly with respect to the developments of the instrumental methods.

John Baldeschwieler
Chairman

ACKNOWLEDGMENTS

The committee is especially grateful to the individuals who made formal presentations to the committee.

At the fourth committee meeting, Richard Richman, Dean Fetterolf, and Frederic Whitehurst, FBI, Department of Justice, gave a presentation on demonstration of Explosive Vapor Detection Techniques; Russell Lease and L. D. "Buck" Goodrich, INEL (Idaho Falls), gave a presentation on the testing facility at Idaho Falls; Jay Stein, Kris Krug and S. David Ellenbogen gave a presentation on Dual Energy X-ray Technique.

At the fifth committee meeting, Steven Smith, IRT Corporation, gave a presentation on Secure 1000 personnel Scanning Systems; and Richard Grogan, Symbol Technologies, Inc., gave a presentation on Bar-Code Technology.

At the sixth committee meeting, Brian Kushner and Sam Fairchild of BDM Corporation, described Inelastic Neutron Scattering (NES) Technology; Donald Watson, Contraband Detection International, talked to the committee about how does one get the various explosive detection techniques into the field? William Davidson, Sciex, reviewed the Condor Detection System; Ken Wood, Barringer Company, gave a presentation on ion mobility spectromety (IMS) for detecting explosives; Martin Annis and Richard Sesnewicz, AS&E, described their company's EDS x-ray imaging system; R. Bruce Miller, Titan/Spectron Company, talked to the committee on energetic x-rays to activate nitrogen in explosives; Joseph Ternes, Veterans Administration, talked to the committee on the use of dogs for explosive detection.

At the seventh committee meeting Donald Greenlee, SAIC, Inc., introduced the subject of simulation of systems architecture, and Mike Smith, demonstrated SAIC's computer-based simulation program; Gloria Bender, American Airlines Decision Technologies, gave a presentation on Use of Simulation Analysis in Evaluating the Operational Impact of Security Systems; David Fine, Thermedics, gave a presentation on Vapor System Integration, Testing, and Evaluation; Tony Feinberg, Stanford University (on leave from OTA), presented and summarized two reports on terrorism; Ruszard Gajewski, and Taiwei Lu, Physical Optics Corporation, presented a talk on Optical Neural Networks Applied to Pattern Recognition; Norman Miller, ScanTech, along with

William Mayo and Richard Mammone, Rutgers University, reported to the committee on Coherent X-Ray Powder Diffraction; John Davies, INEL, gave a presentation on Explosive Vapors Test Results; Paul Schmor, TRIMUF—Canada, gave a presentation on Advanced Systems Development.

At the eighth committee meeting, John Hicks, and James Thurman, FBI, Quantico, VA, gave a presentation on the Pan Am 103 Investigation; John Davies, INEL, presented an update of the FAA IV&V Programs at INEL; Moshen Sanai, SRI, briefed the committee on SRI Container Hardening; Russ Lease, INEL, gave a presentation on Potential Future Terrorist Threats; Tsahi Gozani, SAIC, presented an overview of SAIC's Fast Neutron Analyses (PFNA).

At the ninth and final committee meeting, Paul Bjorkholm, EG&G Astrophysics, briefed the committee on Russian TNA Advances, and updated information on dual energy/dual view x-ray systems; Ann Grow, Sparta, gave a presentation on using fiber-optics and bio-chemical sensors to detect minute quantities of chemicals; Rokaya Al-Ayat, on sabbatical from Los Alamos National Lab to the University of California-Berkeley, presented a Framework for Assessing Operational Impacts of Aviation Security Systems.

The committee would like to express its sincere appreciation to the staff members of the Federal Aviation Administration and Department of Transportation who actively participated in committee meetings.

The chairman thanks each committee member for dedicating time and enthusiasm in preparing this report. The liaison members are particularly thanked for providing valuable support and data throughout the course of this study.

Finally, special thanks go to Robert E. Schafrik, NMAB program officer, and Janice Prisco, project assistant, whose dedicated efforts made possible the production of this report.

CONTENTS

LIST OF TABLES AND FIGURES

DETECTION OF EXPLOSIVES FOR COMMERCIAL AVIATION SECURITY

EXECUTIVE SUMMARY

The threat posed to commercial aviation by small, concealed explosive devices is particularly severe since they are difficult to detect using current techniques and can readily cause tremendous destruction and loss of life. Protecting air travelers from terrorist actions is an essential mission of the Federal Aviation Administration. The FAA plays a critical role in defining the terrorist threat, stimulating the development of explosive detection devices and systems, and in regulating their use. It can issue rules, set minimum performance standards, and set up a system for compliance that will allow private industry and open market competition to play an active role in research, development, manufacturing, and self monitoring. Successful examples of this strategy are the Federal Drug Administration's approach for regulating medical devices, and the FAA's approach for regulating the certification and operation of commercial aircraft. The regulatory framework created by the FAA will influence in a major way the direction of many crucial aspects of the development of explosive detection systems. This would include the resources the FAA will require in order to execute its regulatory duties, the level of investments that private industry would be willing to risk, the pace of performance and cost improvements of deployed systems, and the degree to which air carriers, airport operators, and vendors can cooperate in the common objective of improving U.S. commercial aircraft security.

The committee recommends that the FAA define the regulatory strategy to be used for explosive detection systems. The committee further recommends that the FAA encourage the participation of air carriers, airport operators, and equipment suppliers in the rule-making process.

The United States government has several means to respond to the terrorist threat, which include: deterring the terrorist action; strengthening aircraft against explosions; and detecting and removing an explosive device before it is brought on-board an airplane.[1]

Deterrence assumes that a terrorist would rationally weigh the perceived risks and benefits of the proposed action. From the perspective of a terrorist, risks would include the assessment of the certainty of apprehension and the severity of the punishment. While deterrence may reduce terrorist incidents, it alone is not sufficient to prevent them entirely. Aircraft hardening will increase structural integrity against a given amount of

[1]NMAB-463, *Reducing the Risks of Explosives on Commercial Aircraft*, pages 9-14.

explosive.[2] Hardening, however, adds weight and cost, may require an appreciable time to implement, and may be difficult to accomplish for commuter-sized aircraft.

Deterrence, aircraft hardening, and explosive detection are complementary. Deterrence reduces the number of attempts to penetrate the security system, while a good detection capability acts as a deterrent. Hardening an aircraft increases the minimum quantity of explosives required to cause catastrophic damage. This makes the detection task easier since an explosive detection system would be able to search for larger devices which have a greater likelihood of detection.

The detection of small quantities of explosives is a difficult challenge, but one that is technically feasible. The detection problem is complicated by the following considerations:

- The terrorist has literally dozens of choices of explosives; highly energetic "plastic" explosives are widely available, and thus small, powerful bombs can be inexpensively constructed. Nitrogen-based (nitramine) explosives have three qualities that make them very attractive to a terrorist. They have high energy yields per unit weight, they have small critical diameters (i.e. the small diameter that can sustain a detonation), and they require little or no confinement (i.e. heavy metal walls). However, other types of explosives and devices also pose significant threats.

- Small amounts of explosives have small signatures regardless of the instrumental method used for detection.

- For a given instrument as the detection threshold is lowered, the probability of detection increases, but the probability of false alarms also increases.

- The threat is infrequent among an enormous volume of bags; during the course of a year, only a few devices might be placed in over one billion pieces of baggage.[3]

- A wide variety of items is packed into luggage, presenting a broad spectrum of random background signals to an explosive detection instrument.

- Advanced explosive detection technologies offer considerable promise, but have not yet been demonstrated to be effective, much less cost-effective, in an airport environment.

- An explosive detection system would not be acceptable if it substantially slowed down airport operations, or otherwise adversely affected the flow of passengers through the airport.

---

[2]S. Ashley, "Safety in the Sky: Designing Bomb Retardant Baggage Containers," Mechanical Engineering, January 1992, pp. 81-86.

[3]Report of the President's Commission on Aviation Security and Terrorism, 1990, page 48.

- Fully automated detection equipment can minimize the human role, but cannot eliminate it since judgment will still be required to clear all alarms.

- A sophisticated terrorist can adjust his strategy more quickly than can the opposing security system.

The FAA fulfills an essential function by fostering the continued development of explosive detection devices and their integration into systems. The FAA is funding research and development in relevant technology areas with the expectation that private industry will commercialize the results. **The FAA must establish the standards that explosive detection equipment must meet, as well as the procedures to verify and certify the performance of this equipment.**

The key issues for the FAA regarding explosive detection technology are:
- What can the different detection methods do in principle?
- What can they actually do in practice?
- How can the different methods be best employed to counter the terrorist threat?

This report of the Committee on Commercial Aviation Security addresses the above issues from a detection technology perspective. It discusses and assesses system considerations, testing protocols and performance criteria, and recent explosive detection technology developments.

Key conclusions of the committee are:

- There does not appear at present to be any single detection technology that can, by itself, provide a high probability of detection coupled with a low false alarm rate that will reduce the threat of terrorism at an acceptable cost to airport operations.

- Individual detection devices can be integrated into a system that takes advantage of the strengths of each method. A large range of performance and cost outcomes is possible for a given configuration of detection procedures and devices, depending on the organization of the search strategy, the system architecture and the operational parameters of the individual devices.

- The selection of instrumental methods integrated in an EDS could be based, in part, on the consideration that the vulnerability of one detection technology (EDD) to a potential countermeasure could be compensated by another EDD.

- In order to protect deployed EDS equipment against countermeasure attack, the FAA should work with the airline industry and the EDS equipment suppliers to secure an agreement that the configuration of particular EDS equipment at a particular location will not be made available to the general public. If this cannot be done voluntarily, then appropriate enabling legislation should be sought by the FAA.

- The system architecture or search strategy will vary considerably for each airline and airport situation due to individual terminal designs, utilization rates, and other local factors.

- Within the level of R&D resources available, the technologies chosen by the FAA for advanced development appear to be appropriate.

- The FAA R&D program should not be the sole source of funding for further development of these technologies. A balanced investment strategy that integrates the FAA plan with the program plans of other government agencies and industry would be the most desirable approach.

- The testing of candidate devices should differ significantly from testing of systems ready for deployment. Candidate devices and systems not ready for deployment should undergo parametric testing to verify operational performance at specified levels of statistical confidence. Systems ready for deployment should undergo certification testing in which the system will be judged as pass/fail against a performance specification.

- Accurate, unbiased data on operational parameters for individual detection systems (e.g. probability of detection, false alarm rate, throughput, cost, size, and weight) are essential for rational decisions on the system architecture. Unbiased testing is an essential element of the FAA's Security Technology Program.

  Additional features of a comprehensive certification program would include a process that reviews and encourages cost-effective upgrades to fielded equipment, and provides an atmosphere in which industry will be stimulated to invest in the development of new equipment.

- A general test protocol, which abstracts the most important testing features, insures that all tests consider the same critical factors in a consistent way. A test director, together with a team of experts, could use the general protocol to prepare a detailed test specification tailored for a particular piece of explosive detection equipment without biasing the test.

- The testing of explosive detection devices, in the context of the underlying system architecture, under realistic airport operating conditions, against a FAA-required performance standard must be the keystone of the FAA's certification program. The certification program must include provisions to assure that each explosive detection device in the field will perform over a period of time at least as well as the one that passed the certification test.

- A comprehensive proactive, needs-driven research program can provide options to counter the terrorist threat as it evolves.

SYSTEM CONSIDERATIONS

System considerations include discussions of explosive detection system (EDS) architectures and design issues, and critical issues bearing on the development of computerized simulation tools which could assist in the design and analysis of an operational EDS.

The primary systems engineering challenge is to combine detection devices and procedures in such a way as to achieve high detection probabilities and throughput rates with an acceptable false alarm rate and cost (initial investment and operating costs).

In order for the FAA to address the systems engineering challenge, **the committee recommends that a single organizational entity be responsible for simulation tool development and analysis of explosive detection system architectures.** Computer-based simulation tools and data bases should be developed and maintained under the auspices of the FAA; these tools should incorporate many practical constraints and complex factors arising from airport operating environments which would affect the design and use of explosive detection systems. These tools should be made available to the airline industry, airport operators and designers, and equipment vendors.

TESTING PROTOCOLS AND PERFORMANCE CRITERIA

Effective, unbiased testing of explosive detection equipment is an essential element of the FAA's Security Technology program. The development of standard test protocols and performance criteria, and the FAA's role in testing are key ingredients of airport security regulatory process.

Because of the importance of testing, it is **essential that the organization charged with conducting such tests be independent of and insulated from potential pressures internal and external to the FAA.** Furthermore, this activity must be conducted by an organization with established expertise in testing matters using specified protocols.

An important distinction is made regarding the categorization of the equipment being tested. An Explosive Detection Device (EDD), which employs one particular instrumental method to detect the presence of explosive material, would be tested to verify that its performance matches the data provided by the manufacturer. On the other hand, an Explosive Detection System (EDS), which could be composed on one or more integrated EDDs, would be certified by the FAA as meeting the operational standards.[4] **A comprehensive EDS certification program which includes but is not limited to certification testing, is strongly recommended by the committee.**

GENERAL TEST PROTOCOL FOR BULK DETECTION

A general test protocol containing the framework of significant testing considerations required to design a test plan, conduct the tests, and analyze the test data and evaluate bulk detection equipment was developed in

[4] Parametric testing could be performed on EDS equipment to fully characterize the operational characteristics prior to the formal certification process.

consultation with the committee.[5] It is a guide in designing detailed verification and certification test data analysis plans. This protocol clearly delineates the differences between verification testing and certification testing, elaborates on the composition of the standard bag set, requires documentation of the rationale for deviations from the protocol, and includes provisions for updating as additional test experience is gained.

STATUS OF VAPOR AND PARTICLE DETECTION TEST PROTOCOL

The committee was asked to develop a generic test protocol for vapor detector instruments. By their very nature, the results from vapor detectors are inferential; i.e when a vapor is detected the presence of the bulk explosive (threat) associated with the test object must be inferred. From vapor detection alone, nothing can be said about the amount of explosive present. For instance, vapor in sufficient quantity to cause current detectors to alarm can come from crumbs of explosive material. Conversely, amounts of explosive ten times the lethal threat quantity, if properly encapsulated, would not necessarily cause these detectors to alarm because there would be no vapor to detect. Aside from the special cases of complete containment or the availability of particulate material, the amount of vapor available for sensing from a given quantity of a concealed explosive is not known. At this time, there is no certified vapor generator available to produce a known small quantity of an explosive's vapor, and there is no reference instrument available to check these generators or the background contamination at low levels.

Until the role of vapor detection in the explosive detection system can be specified and the level of detection necessary to fulfill successfully this role is quantified, a generic test protocol cannot be prepared.

The committee did provide informal input to a group at Idaho National Engineering Laboratory (INEL) preparing a preliminary detailed test specification for developmental testing of candidate vapor detection devices. This specification is still evolving as the various test procedures are reduced to practice. **The committee recommends that, if vapor detection devices are to be certified, a test facility must be maintained that has standard vapor generators and a reference instrument. The facility will have to be free from contamination and capable of remaining in that condition after the tests.**

RECENT TECHNOLOGY DEVELOPMENTS

The committee reviewed and updated the status of specific technologies and devices currently under development for explosive detection. Many of these technical approaches were the result of investments by the FAA's research and development (R&D) program or by industry encouraged by the FAA's

---

[5] In this context, bulk detection refers to the sensing of some physical or chemical property of the solid phase of an explosive.

active interest in soliciting their ideas. **The committee did not find any significant approach that had not been explored by the FAA through direct funding, by monitoring, or by a combination of funding and monitoring.**

Broad investment strategy priorities are suggested below for various technologies. Table ES-1 is a short summary of each detection technology.

The assignment of priorities was guided by the following considerations:

- High priority projects directly support devices and systems which could be deployed within the next several years, or support efforts with a longer lead time that could have a high future payoff.

- Medium priority projects, if successful, could result in an improvement of the next generation of EDS, or could have a payoff of intermediate value.

- Low priority projects were considered to be technically high risk or to have a low payoff, which should be pursued only as funds allow.

For the purpose of evaluating the stage of development for the various instrumental technologies, the following life cycle phases were used:

- Concept: the concept is clearly described, proof-of-principle calculations performed, and drawings of laboratory experimental set-up completed.

- Demonstration of Principle: laboratory apparatus assembled, signal-tonoise measurements completed, and detection of pure standards and interferrants tested.

- Engineering Prototype: detection module integrated with other devices and sub-systems, testing of key operational parameters completed, and physical size and facility requirements of a fully capable device defined.

- Deployable Device: specifications, operational tests, manufacturing and assembly methods, configuration and software finalized; pricing established; available for integration into a qualified explosive detection system.

Following is a summary of the instrumental technologies, which have been updated from the previous report:[1]

1. Thermal Neutron Activation (TNA)

TNA provides an important detection capability against larger quantities of explosives. Although the probability of detection decreases and probability of a false alarm increases for smaller quantities of explosives against the background of nitrogen containing materials in passenger luggage,

TABLE ES.1. Summary of Explosive Detection Devices (EDD)

EDD Technique	Principle Of Operation	Characteristic Detected	Advantages
Thermal Neutron Activation (TNA)	Low energy neutrons captured by nitrogen atoms, resulting deexcitation produces characteristic gamma rays.	Nitrogen content.	Many highly energetic explosives have high nitrogen content. Operational experience with some pre-production equipment. Can be automated; regions of high nitrogen content identified.
Elastic Neutron Scattering	Monoenergetic neutron source used to scan objects; elastically back-scattered neutrons are detected.	Carbon, nitrogen, oxygen content calculated from neutron energy loss.	Measures quantities of light elements. Suspect location identified.
Pulsed Fast Neutron Activation (PFNA)	Fast pulses of neutron beams used to excite characteristic gamma rays.	Provides carbon, nitrogen, and oxygen compositional information.	Measures quantities of light elements present. Determines position and depth of suspect material.
Photon Activation	A powerful electron linear accelerator produces bremsstahlung x-rays, which in turn produce a radioactive isotope of nitrogen when encountering nitrogen atoms. The resultant nitrogen isotope has a 10 minute half life, and decays by emitting a positron.	Nitrogen content.	Builds on experience with existing medical PET devices. Excellent spatial resolution possible.
Nuclear Resonant Absorption (NRA)	Proton beam bombards a target to produce high energy gamma rays which preferentially excite nitrogen atoms.	Nitrogen content.	Able to penetrate shielding around an explosive. High sensitivity combined with good spatial resolution for detecting nitrogenous material.
Fast Neutron Associated Particle (FNAP)	High energy neutrons produced from a deuterium-tritium reaction are ejected in known directions relative to an ejected associated alpha particles. These fast neutrons activate nuclei by inelastic scattering which results in emission of characteristic gamma rays. Correlation of the direction of the gamma rays with the alpha particles yields directionality.	Relative amounts of carbon, nitrogen, and oxygen.	Provides elemental compositions and locations of explosives. Uses highly penetrating neutrons.
Dual Energy X-Ray	Alternating x-ray beams of high and low energy levels produce two different images due to differences between the photoelectric and Compton attenuation coefficients of the different elements. By comparing the images, the areas with light elements can be identified.	Average atomic number, density, and shape.	Extension of a medical device. Provides narrow resolution in average atomic number which allows identification of materials.
Backscatter Analysis X-Ray	Compares normal x-ray transmission image with a Compton backscatter image. By comparing the images, the areas with light elements can be identified.	Average atomic number, density, and shape.	Uses off-the-shelf technology. Extension of x-ray technology already familiar to airport security. Can be readily available.
Extremely Low-Dose X-Ray	Same as backscatter analysis x-ray but at a lower x-ray intensity level.	Average atomic number, density, and shape.	Can be used on people. Extension of available medical devices. Can be readily available.
Coherent X-Ray Scattering	X-ray diffraction.	Crystal structure.	Depends on unique crystal structure of the explosive.
Dual Energy X-Ray Computed Tomography (CT)	A CT image is a map of the x-ray attenuation coefficient in each voxel. Attenuation depends on density and composition. Differences between photoelectric and Compton attenuation coefficients at two different energy levels are used to solve for density and composition.	Shape, atomic number, and density.	Similar to medical CAT scanner. Produces true cross-section slices. Suspect areas can be imaged in greater detail. Can be readily available.
Vapor/Particle Detection Devices	Devices employ a variety of methods, including: Gas Chromatography; Chemical Luminescence; Mass Spectrometry.	Volatility, molecular weight, and electron affinity.	Non-invasive method. Can be used on people. Experience in other applications. Commercially available equipment.
Dogs	Not known for certain. May sense chemical vapor from the explosive itself or some other odor associated with the explosive; or may sense particulate components from the explosive.	Volatility and possibly other characteristics.	Non-invasive. Very mobile. Can cover wide area rapidly. Can be used on people.

Limitations	Critical Issue(s)	Development Stage	Committee Recommendation
Only detects nitrogen; large size; sensitivity degrades as explosive quantity decreases; uses radioactive source. Cannot be used on people.	Sensitivity against background and interferents.	Pre-production prototypes deployed at several international terminals.	High priority for continuing to refine the devices already deployed until further efforts are assessed to have low marginal benefit.
Shielding required; background scattering noise reduces sensitivity. Cannot be used on people.	Requires accelerator to produce 1 MeV to 5 MeV proton beam. Must demonstrate time-of-flight measurement capability.	Laboratory demonstration of principle.	Low priority for the development of an engineering prototype.
Shielding required. Large size. Durability unknown. Cannot be used on people.	Engineering issues related to the accelerator (must produce approximately 7.5 MeV neutron beam), target, scanning system, and detectors.	Engineering prototype.	Medium priority for the testing of the critical elements of an engineering prototype.
Only detects nitrogen. Susceptible to background interference from other elements. Would destroy any radiation-sensitive material in baggage (e.g. film). Cannot be used on people.	Engineering issues related to the accelerator (13.5 MeV).	Laboratory demonstration of principle.	Low priority for the development of an engineering prototype.
Only detects nitrogen. Durability unknown. Cannot be used on people.	Development of high current accelerator. Long life multi-layered carbon target. Data acquisition and analysis system.	Laboratory demonstration of principle.	Low priority for the development of an engineering prototype. High priority for development of accelerator technology.
Shielding required. Severe background problem. Requires nuclear accelerator. Requires multiplexed alpha particle detection system. Cannot be used on people.	Handling of radioactive tritium target. Reliability of accelerator. Design and operation of detector system. Processing time required for the computerized algorithms.	Laboratory demonstration of principle.	Low priority for supporting the development of a prototype.
Looks for average atomic number, not a specific atomic number. Cannot be used on people.	Complex computer analysis required.	Engineering prototype.	High priority for the testing of existing engineering prototypes.
May have a high false alarm rate due to difficulty of performing computer analysis. Cannot be used on people.	Effectiveness and speed of computer analysis.	Engineering prototype.	Medium priority for testing of existing engineering prototypes.
Image may be confusing due to diversity of articles carried by travelers.	Public acceptance of even a low radiation dose. False alarm rate.	Engineering prototype.	High priority testing of existing engineering prototypes.
Cannot be used on people.	Scan speed. Verification of unique explosive crystallinity "signatures."	Concept.	High priority for demonstration of principle.
Slow speed may require much greater computer power. No experience to date in scanning actual passenger bags. Cannot be used on people.	Speed of the analysis algorithm.	Engineering prototype.	High priority for performance testing of improved prototypes.
Very low vapor pressure of plastic explosives results in femtograms of available vapor. Sample collection step is critical. Does not give a direct measure of the amount of explosive present.	Demonstrated effectiveness in detecting high performance explosives in luggage. May be more effective as particle detectors since much more sample materials is available for analysis.	Varies. Some units are commercially available.	High priority for determining quantity of vapor associated with explosives in baggage and carried on persons; high priority for the development of particulate detection standards; and high priority for well-controlled testing of deployable devices with particulate detection standards.
Reliability over a period of time. Performance difficult to quantify.	Determining absolute detection threshold for explosive vapors, and what key factors influence a dog's detection performance.	Operational.	High priority for performance testing of dogs.

TNA still provides a significant detection capability for smaller quantities of explosives in selected items with an inherently low nitrogen background, such as electronic devices and smaller items of carry-on baggage. TNA is likely to be an important element in a larger architecture of detectors. The committee recommends continued support for the refinement of this deployable device at a high priority level until an alternate explosive detection device of greatly improved performance becomes available, or until further improvements result in only small marginal gains.

2. Automated Neutron Source Accelerator

Although existing TNA devices utilize a radioactive ^{252}Cf isotope as a source of thermal neutrons, the source might be made more compact, require less radiation shielding and be less expensive if a small, completely automated particle accelerator were developed for the neutron source. In fact, all of the proposed nuclear-based explosive detection devices, with the exception of TNA, are dependent on the development of specialized particle accelerators. Therefore, development of this type of accelerator is assigned a high priority by the committee.

3. Elastic Neutron Scattering

This technology in principle can detect specific elemental components of an explosive by measuring the energy loss of elastically scattered neutrons. However, no tests have yet been made using realistic baggage while imposing the shielding conditions necessary at an airport. Much more research and development is needed before a judgment can be made regarding the practicality of this technology as an explosive detection device. The committee gives the development of an engineering prototype device utilizing this technology low priority.

4. Pulsed Fast Neutron Activation

In principle, PFNA is the most promising of the nuclear technologies, since it can be used to identify the different atoms that make up an explosive -- e.g. carbon, nitrogen and oxygen -- and identify the explosive by its composition, quantity, and location. The pulsed fast neutrons make possible fast correlation analysis that then allows for three-dimensional construction of the relative location of the explosive materials within the suitcase on a single pass. This technology is currently under development for use as a device for scanning entire containers for explosives before allowing their entry into the tunnel under the English Channel. However, the technology suffers from potentially high cost and complexity. The committee recommends medium priority for the testing of an engineering prototype.

5. Photon Activation

This approach activates nuclei with high-energy photons produced by a powerful electron accelerator. The resulting radioactive nitrogen and other nuclei then decay by positron emission, allowing for both detection and location by positron tomography methods. Significant safety and passenger acceptance issues involving the high level of gamma irradiation, as well as cost and complexity of the tomography detector system, make this technology seem less promising than some of the alternative nuclear methods. Low priority for development of an engineering prototype is suggested by the committee.

6. Nuclear Resonant Absorption

This technology utilizes a high energy gamma ray that is strongly absorbed (resonant) by nitrogen and nothing else. A very high current accelerator is required along with specially prepared targets and possibly detectors as well. Three different organizations are working on this technology; however, the engineering problems are formidable. Although it would be useful to expend effort on accelerator development, development of an engineering prototype is accorded low priority by the committee.

7. Fast Neutron Associated Particle (FNAP)

This method utilizes the nuclear reaction of two hydrogen isotopes, deuterium and tritium, that results in a 14 MeV neutron and an associated alpha particle. The 14 MeV neutron and alpha particle move in opposite directions, so that detection of the alpha particle's direction with respect to the target or nuclear reaction point determines the direction of the neutron. The alpha particle detection also provides a start signal for a neutron time-of-flight measurement for any detected deexcitation gamma rays, which can then be used for elemental identification and three dimensional imaging. Although this method has potential as an explosive detection device, the engineering problems are formidable and a working prototype has yet to be demonstrated. A low priority is recommended by the committee for the support of prototype development.

8. Dual Energy X-Ray Systems

This promising technology uses alternating beams of high and low energy-x-rays to produce two different images. This technique relies on the differences between the photoelectric and Compton attenuation coefficients of elements for the two different x-ray energies. Heavy elements are more effective at absorbing high-energy x-rays while light elements are more effective at scattering low energy x-ray beams. The differences in signal between the two x-ray beams can be used, after suitable analysis, to compute the average atomic number of an object.

The results to date are impressive. Near-term implementation at reasonable cost appears feasible. The committee recommends that testing of existing engineering prototype devices for deployment be given high priority.

9. Backscatter X-Ray

This x-ray technique makes use of two images: the normal transmission image which shows areas of photoelectric absorption by the high atomic number elements, and backscatter image due to compton scattering. Since compton scattering is relatively independent of atomic number unlike photoelectric absorption, the comparison of the backscatter image to the transmitted image, allows the average atomic number of various areas to be computed.

The ability of this technique to discriminate explosives from other "low-Z" materials such as paper has not been demonstrated. However, the method, if successful, would be relatively inexpensive. The committee suggests that testing of the existing engineering prototypes for feasibility be given medium priority.

10. Extremely Low-Dose X-Ray Devices for Searching Passengers

The performance of these devices, which use x-ray backscattering, is very impressive. The radiation risks are reported to be negligible. However, passenger acceptance will be a major issue, as will false alarms. Since almost all objects carried on a person will be readily detected, discrimination between threatening and non-threatening objects will be a problem. At present, these devices provide the only demonstrated capability for detecting explosives carried on a person. The committee recommends that testing of existing engineering prototype units for deployment be given high priority.

11. Coherent X-Ray Scattering

Coherent x-ray scattering is similar to conventional x-ray diffraction in that Bragg's Law is used to compute crystal structure and lattice spacings.

The powder diffraction pattern can then be compared to known x-ray powder diffraction patterns of explosives of interest.

The capability to identify specific explosive compounds from their powder diffraction pattern is impressive. This technology appears extremely promising and has significant long-range potential. Demonstration of principle should be given high priority.

12. Dual Energy X-Ray Computed Tomography (CT)

A computed tomography image is a map of the x-ray attenuation coefficient in each volume element of an object. Typically, the object is imaged one cross-section at a time, and a three-dimensional image reconstructed from each slice. The x-ray attenuation coefficient depends on both the density and composition of an object. Dual energy imaging (as noted above) provides enough additional information to determine the individual contributions of density and composition to the x-ray attenuation in each volume element.

Dual energy CT has the potential to determine the shape, atomic number, and density of objects in luggage. It holds significant promise for a detailed analysis of suspect pieces identified by other screening methods. A high priority is suggested for testing engineering prototypes.

13. Vapor Detection Devices

The major virtue of vapor detection devices is their non-invasive nature and applicability to screening passengers. But, vapor detection suffers serious problems arising from the lack of understanding of the mechanisms involved in the evolution of the vapor, and potentially high false alarm rate. Therefore, at the present time, the committee concludes that vapor detection devices geared to sensing extremely small concentrations (femtogram level) of explosive vapor are unsuited for use in airport terminals as a primary explosive detection method. The committee recommends the FAA focus a research activity toward determining quantitatively the amount of vapor available for detection in baggage and passenger scenarios.

However, current vapor detectors, if suitably configured, can readily detect particulate plastic explosives. The amount of material available for detection from particles can be more than a million times greater than that available from vapors. Particulate detection, rather than vapor detection, represents a technology that could be successfully deployed very quickly. Testing standards and airport calibration and testing would be relatively straightforward for particulate detection and the devices would not be challenged to work at their absolute limit of detection. However, no particulate standards are currently available. The FAA should assign a high priority to the development of particulate detection standards.

14. Dogs

Dogs potentially provide a viable capability, although they have limitations in search time capability and performance retention over long time periods. Controlled tests of dog performance equivalent to those used with explosive vapor devices will be fundamental to developing a rationale for their use. The committee recommends that such testing be given high priority.

15. Bar Coding

Positive matching of each piece of baggage to a passenger can be an effective terrorist deterrent. Unaccompanied bags would be treated as suspicious and would not be brought on-board an aircraft without a thorough examination. The application of linear bar code technology offers significant improvement over existing manual-intensive approaches. Currently several different bar code technologies are being applied by air carriers. The committee recommends that international standardization of bar coding symbology be given a high priority to allow bag tracking among all carriers world-wide.

16. Pattern Recognition

This emerging technology area offers the potential to automate the interpretation of data from various detection instruments. Implementation of pattern recognition algorithms can help automate the interpretation of results from the instruments, reducing the dependence on human judgment in recognizing explosive materials in scanned luggage. Various technical approaches are possible, including different implementations of neural networks. The committee recommends a medium R&D priority to determine the most effective pattern recognition approaches.

SUMMARY OF TECHNOLOGY RECOMMENDATIONS

The following efforts are recommended with a **high priority**:

- Continued testing of Thermal Neutron Activation units at airports, and refinement of the method until a determination is made that little additional benefit is being gained.
- Development of an Automated Neutron Source Accelerator.
- Performance testing of a Dual Energy X-ray prototype.
- Demonstration of principle of Coherent X-ray Scattering for explosive material detection.
- Performance testing of Dual Energy X-ray Computed Tomography engineering prototypes.
- Performance testing of Extremely Low Dose X-ray prototypes.
- Statistically-Valid Testing of Deployable Vapor Detection equipment using a test protocol.
- Development of particulate detection standards, and testing of deployed devices using the standards.
- Performance testing of Dogs.
- International standardization of Bar Coding for checked luggage.

The following efforts are recommended with a **medium priority**:

- Performance testing of Pulsed Fast Neutron Activation prototypes.
- Performance testing of Backscatter X-ray prototypes.
- Demonstration of proof of concept of Pattern Recognition approaches.

The following efforts are recommended with a **low priority**:

- Development of an Elastic Neutron Scattering engineering prototype.
- Development of a Photon Activation engineering prototype.
- Development of a Nuclear Resonant Absorption engineering prototype.
- Development of a Fast Neutron Associated Particle engineering prototype.

1

SYSTEMS CONSIDERATIONS

A. INTRODUCTION

The technology for detecting explosive material in checked baggage is continuing to advance. Several instrumental methods have demonstrated, or are close to demonstrating, at least some operational capability. These explosive detection devices (EDDs) will become essential building blocks for an explosive detection system (EDS) that could reasonably be installed in airport terminals and operated day in and day out at predictable performance levels.

Countermeasures can degrade system performance, perhaps fatally. Considerations of countermeasures are not to be addressed in this report, but must be taken into account during the design and implementation of an EDS.

This chapter discusses two important considerations for an explosive detection system.

B. EDS ARCHITECTURE AND SEARCH STRATEGY

Explosive Detection Systems cannot be the sole means used to counter the threat posed to commercial aviation by small, concealed, highly energetic explosive devices. Within the larger frame of reference, this threat can be responded to by a wide range of activities involving deterrence, aircraft hardening, and detecting and removing concealed explosive devices. Within the airport environment, the definition of the role of explosive detection in the overall security program provides important inputs for defining the internal EDS architecture. The most important of these inputs include: (a) amount and morphology of each type of explosive material required to be detected (i.e. the threat); (b) specified degree of confidence in finding the explosive material; (c) maximum acceptable false alarm rate; (d) processing speed capability as reflected by the required through-put parameters of passenger and baggage volume, rate and type; (e) method for resolving alarms; (f) likely countermeasures; and (g) provision for future change as advancements occur and requirements evolve.

The EDS systems engineering challenge is to combine sensors and procedures in such a way as to meet the detection probability and throughput rate requirements with acceptable false alarm rate and cost without being vulnerable to defeat by likely countermeasures. These requirements will strongly influence the architecture of an explosive detection system.[1] To date, no single instrumental method has been shown to be a "silver bullet" which can satisfy all of the requirements. Therefore, it is highly likely that two or more complementary instrumental methods (i.e. EDDs) will be required. The following discussion summarizes the key considerations relating to the internal logic of integrating the responses from multiple detectors which sense different physical characteristics so that the responses are independent of one another.

The trade-off between the acceptable system level of detection probability (when there is an explosive) and the false-alarm rate (when there is no explosive) is crucial to the system architecture; specific trade-off decisions cannot be made without data on individual instruments. Thus, statistically sound, detailed parametric performance data for individual EDDs must be available for the analyses. Chapter 2 discusses key considerations for obtaining such data.

It is generally assumed that all baggage made available for EDS testing will indeed be examined by the EDS. However, in the analogous industrial context of monitoring the quality of manufactured products and seeking to find defective items, one can easily imagine inspection sampling plans in which some portion of products bypass examination. Such a plan conceivably could be cost effective when the financial costs of sampling and failing to detect a flawed item are well understood and when the latter penalty is relatively low (e.g., routine warranty costs and no possibility of excessive litigation). But, in the screening for an explosive in baggage the actual and perceived costs of failing to detect an explosive are no doubt extremely high, and exact values cannot be definitively quantified. One can imagine the public outcry that would arise if an airplane is destroyed by an explosive that could have been readily detected by the EDS, but was instead randomly designated to bypass testing.

Moreover, any potential gain in efficiency attained via subsampling is necessarily limited by the prescribed high value for the EDS probability of detection, P_D. For example, if the required P_D threshold is 0.95 then clearly no more than five percent of the baggage can be permitted to bypass inspection (even when the EDS is infallible).

All alarms will have to be cleared. At least three alternatives are available: re-run the bag through the unit; operator interpretation of the EDS-generated image of the bag; or hand searching the suspect bag. The impact of each of these alternatives could be examined with the simulation tools. Additional complexity in the search strategy can be envisioned. For instance, the results from passenger profiling as part of a pre-screening operation could be used to cue the detector threshold levels in the instruments (i.e. lower the detection level for the higher "risk" passengers). In some

[1]For a discussion of the entire airport security system architecture for detection explosives see NMAB-463, pp. 14-16, and OTA-511, pp 71-75. Here, we are only concerned with system architecture relating to the logical integration of EDDs.

instances, the viewing angle of luggage could be changed to facilitate a particular aspect of the inspection process based on a "clue."

In order to protect deployed EDS equipment against countermeasure attack, the FAA should work with the airline industry and the EDS equipment suppliers to secure an agreement that the configuration of particular EDS equipment at a particular location will not be made available to the general public. If this cannot be done voluntarily, then appropriate enabling legislation should be sought by the FAA.

C. SIMULATION OF EDS DESIGNS

Computer simulation can provide very powerful analysis tools in developing design alternatives, trade-off strategies, and expected behavior of complex systems. The application of simulation also imposes a solution framework on the problem, which itself can be a very useful way to define and structure the problem in an iterative fashion. Simulations which can aid the assessment of incorporating various explosive detection strategies into current airport terminal operations will be of value to the FAA, carriers, and airport operators.

Therefore, the concept of using simulation for designing an airport explosive detection system is appealing. Once the tools are developed, and the supporting databases are fully populated, the consequences of changing assumptions, system input parameters, system details, etc., can be quickly explored. The result can be a set of well-analyzed alternative EDS configurations with the appropriate advantages and disadvantages described.

Simulation, however, is not a panacea. The development of good simulation models is an expensive, time consuming effort which requires the dedication of high caliber experts. Simulation can give very good, or very bad, results, depending on how it is used and how faithfully the underlying simulation models represent the "real world." The results of simulation must, by their nature, be imprecise, but there may be a tendency to attribute greater precision to the numerical results than is warranted. In the following discussion, no attempt has been made to characterize the difficulty and potential pitfalls of developing and applying different simulation models. The discussion focuses on describing WHAT is required. But the assumption should not be made that the process of developing and validating these models is either easy or quick. **The committee recommends that the FAA initiate work on developing these simulation tools as soon as possible.**

The overall requirements for a robust set of simulation tools fall into two categories:

- analysis of the internal EDS architecture; i.e. providing support to accomplish the analysis mentioned in the previous section; and

- selection and integration of the best EDS choice for a specific airport environment; i.e. providing insight into the most cost-effective combination of specific FAA-certified EDS configurations for screening all passengers and their baggage.

The previous section described the general input requirements for the simulation of an EDS architecture. As mentioned, no single explosive detection device is expected to meet all the performance requirements. For instance, a combination of devices involving x-ray analysis to delineate a suspicious object and nuclear methods to identify high nitrogen concentrations has been suggested as potentially offering a satisfactory technological approach at present.[2]

The FAA will collect performance data for the various explosive detection devices obtained from independent testing organizations to the extent it is available. The data should include the sensitivity of the detection methods to different types of explosives, different quantities and different shapes. Since some of these data may be classified, the simulation model should be able to access classified information without retaining it so that its products could remain unclassified. But, depending on FAA's security classification guidelines, the output of the model may sometimes have to be classified .

Simulation runs should be made over a range of worst case to best case parameters (e.g. varying the amount of explosive material in a bag at different through-put rates) for each device in order to determine the realistic performance from various device combinations and operating sequences.

The following airport-specific input requirements to the simulation model would include: (a) airport operations scenario; (b) available location and amount of space for the EDS units; (c) unique constraints of the air terminal; and (d) cost.

With the appropriate simulation model, various EDS configurations for checked and carry-on baggage, and for passengers, could be simulated in the context of a specific airport, and the resulting performance assessed under varying conditions. Outputs from the computer model would include: number of EDS units required for the passenger load, overall system cost, detection effectiveness, false alarm rate, peak and average throughput, and specific impacts on airport and airline operations including environmental effects. This information would assist in developing alternatives and in decision-making for deploying these systems into existing terminals. This simulation capability would be extremely useful for future airport terminal design where the number of constraints would be small.

The simulation tools should be available to the FAA, policy makers, air carriers, airport operators, airport designers, and manufacturers of explosive detection equipment to provide guidance for making rational decisions for deployment of inspection hardware, personnel, etc., into an effective overall plan within acceptable air carrier and airport operational requirements.

---

[2]OTA-511, *Technology Against Terrorism-Structuring Security*, Jan 92, pages 99-100.

Study Airports

The simulation model should be constructed in modules so that the various levels of analysis can be performed in a loosely coupled way. One of the modules should be capable of representing a wide variety of existing terminal configurations, baggage handling and operating systems. The module should be flexible enough to handle unusual baggage check-in configurations. This module should provide an analysis of passenger and baggage screening measures.

In order to gather the necessary airport-specific data, a number of terminal and airport configurations should be studied and used to develop a common architecture, or generalized framework, for airport operations that would affect the explosive detection system. This framework could then be specialized for particular airports of interest.

A detailed functional assessment and database development for airport environments should be accomplished. The committee suggests that as many as 12 existing U.S. airport terminals be analyzed for the purpose of developing a framework which contains the key functional aspects of an airport that an EDS would affect, and designing a database which would contain the relevant details of the airport. As an example of the level of detail required, the framework should be comprehensive enough to model the space required to operate checked baggage inspection systems within various ticketing and check-in counter baggage handling configurations and carry-on baggage inspection systems. This information would form the basis for developing realistic simulations of EDS operations in widely diverse airport and air carrier operating environments so that the most efficient use of space can be determined, as well as the best way to integrate an EDS with the existing baggage handling systems.

A primary goal of the modeling activity is to incorporate a sufficient range of space and operational circumstances to provide analytical support for airport architects who can either modify existing space or design space to accommodate the new baggage screening equipment. The architects want to minimize system costs and the number of screening systems and to avoid unnecessarily compromising the efficiency of the terminal baggage check-in operations, passenger services, and/or overall terminal service operations.

Care should be taken to select an appropriate cross-section of terminal buildings that vary in size, number of air carriers, number of common checked baggage systems, passengers per peak hour to be served, operating power, other operational environments, structural support systems, space dimensional needs, distances to baggage check-in areas, and bags per hour (peak and average).

The criteria recommended for selection of the 12 different airport terminals are summarized in Table 1.

- In the worst case categories, two of the airports should be among the five busiest airports in the United States with domestic and international operations, and with substantial complexity of operational and space environments.

▪ Two terminals should be chosen from those designated as having specialized operations (e.g. FAA designated Category X) due to their locations.

▪ Two should include small (phase three) airport terminals having differing baggage check-in facilities and operating conditions.

TABLE 1. Airport Terminal Site Selection Factors

Airport Categories	Number	Comment*
Among the five busiest with both domestic and international operations	2	Substantial complexity of operational and space environments
Specialized operations due to significance of the location	2	Each with different baggage check-in facilities and operating conditions
Small terminals	2	Each with different baggage check-in facilities and operating conditions
Extreme weather locations	3	•Extreme cold •Extreme humidity •Extreme dry heat
Service a mix of international and domestic passengers •New terminal (<5 years old) •Older terminal (>20 years old)	2	Each airport should have similar passenger mix
Centralized terminal operation	1	Serves more than one air carrier

*At least one terminal should be planning renovation/expansion.

▪ One should involve a terminal operation that functions under extreme cold weather conditions; one that functions under extreme moist heat conditions; and one other that operates under extreme dry heat conditions.

▪ One terminal complex should have a mix of international and domestic passengers and be within five years of age. Another should have a similar passenger mix, but be older, (i.e., 20 or more years old). For the case of a large international airport, distinction between domestic and international air carrier operators and screening procedures for passengers' checked baggage and carry-on baggage should be made in the data.

- A terminal complex with one or more centralized operations within a single operational zone serving more than one carrier should be selected.

- Selection of the candidate terminals should include those facilities being considered for remodeling and expansion.

The effort to gather existing and new data will require site visits to each of the selected terminal sites. Discussions with airport, air carrier, and FAA representatives will also be required. It can be anticipated that degrees of cooperation and sharing of available data and information about the checked baggage systems and baggage inspection processes of selected airports will vary, and that gaps in the data may have to be filled by visiting additional airports.

Model Validation and Analysis

The simulation model should be validated incrementally as key modules are completed. The modules could be validated by comparing the results of the model for a particular airport (using current airport operations) to actual performance at that airport. Performance factors such as passenger and baggage flow rates associated with peak hour, average day, and peak month actuals versus simulated should be compared. Results from EDD and EDS demonstration tests, if well-controlled and statistically valid, could also aid validation of the models.

The modeling should allow analysis of the implications on space requirements and air carrier operations associated with new federal regulations pertaining to checked baggage explosive detection and expansion of FAA screening requirements having to do with passenger and baggage explosive detection and screening. For example, it would be of interest to simulate the impact of the following scenarios:

- the explosive detection screening of all checked baggage for all international departures;

- the explosive detection screening of all checked baggage for all domestic departures;

- application of positive passenger/baggage matching procedures for use throughout the checked baggage explosive detection process and up to the time of aircraft departure;

- elimination of carry-on baggage, other than essential items, for departures;

- opening bags for insertion of test probes; and

- restricting carry-on luggage size and contents (segregation into bags containing only clothing versus metal and other objects).

The expected impact of the implementation of the foregoing scenarios could be simulated for at least some of the selected study terminal sites.

The objective of these simulations would be to identify the extent and cost associated with appropriate solutions. The analyses should be capable of being expanded to include the identification and evaluation of alternative conceptual layouts and screening equipment to fit into available space. The simulation should include an assessment of costs required to implement each potential alternative. Results of these efforts could serve as a basis for identifying guidelines for the appropriate deployment of equipment and development of responsive passenger/baggage flow patterns to meet potential alternative levels of passenger and baggage requirements. Output of the simulations should be presented in a manner that will facilitate a clear understanding of the findings, including dynamic graphic displays. The output could also be presented so that relationships and activity levels should be capable of further study and evaluation. Findings of the study should highlight the advantages and disadvantages associated with certain design solutions under differing levels of activity, including generalized conceptual equipment deployment and flow patterns, and the resulting sensitivity and specificity for explosive detection.

Key Simulation Products

Potential outputs resulting from a comprehensive simulation modeling activity are:

- confidence level that an EDS would meet the FAA-mandated requirements, including threat detection, baggage processing rate, and false alarm rate, in the context of overall airport operations;

- number of EDS required to service passenger traffic;

- costs and operating impacts on air carriers and airports;

- optimum locations for screening equipment;

- Airport Terminal Architectural Considerations
  - space requirements for baggage processing;
  - spatial relationships within the baggage and passenger handling areas of the terminals between explosive detection and other activity centers within the terminal itself;

  - location(s) of entrance and exit areas for passengers;
  - structural loading implications;
  - partition locations;
  - stacking areas for baggage make-up and passenger handling; and

- other considerations relating to the passenger and baggage screening processes.

2

TESTING PROTOCOLS AND PERFORMANCE CRITERIA

A. INTRODUCTION

Effective, unbiased testing of explosive detection equipment is an essential element of the FAA's Security Technology Program. It is only through a well planned test and evaluation (T&E) program that realistic evaluations and demonstrations of the ability of equipment to satisfy FAA-mandated explosive detection requirements can be assured. By the time a deployment decision is made, T&E should encompass all of the elements of the explosive detection system: equipment, software, facilities, personnel, and test procedures.

The various types of testing mirror the development phases. Developmental testing occurs primarily during the engineering prototype phase. Its goals include: verification of the performance characteristics of a particular device; input into the engineering design and development process; demonstration that design risks are manageable; estimation of utility; evaluation of compatibility and interoperability with existing/planned systems; and assurance that the equipment/system is ready to proceed to testing in the operational environment. Additional limited developmental testing can be conducted after devices have been fielded. Typical reasons include: verification of the effectiveness of product improvements and demonstration of the adequacy of redesigns to solve field or production problems.

Testing of full configuration equipment in an operational environment can be conducted to characterize a device, or to qualify a system for certification testing. The goals are to assess functional characteristics and to demonstrate the equipment's operational effectiveness and suitability. The items tested must represent production models so that a valid assessment can be made. This testing is usually conducted by a group that is separate and independent from both the developer and the user.

Field level verification/validation testing is normally conducted at operational sites on deployed equipment. The goal is to measure the continued level of effectiveness of the equipment under normal conditions, using regular operators and maintenance personnel.

This report addresses generic test protocols required to verify functional characteristics of EDD and EDS equipment, and certification testing of an EDS. It does not address specific protocols required for developmental testing, field level verification/validation testing, or for Explosive Detection Operation that includes activities other than those accomplished by the EDS, such as profile analyses and terrorist intelligence. Further, it does not address total Airport Security Operations, e.g., access control systems.

B. ROLE OF TESTING IN THE FAA REGULATORY PROCESS

The FAA plays a complex role both in stimulating the development of explosive detection devices and systems, and in regulating their use. A well planned test and evaluation (T&E) program should provide realistic evaluations and demonstrations of the capability of equipment. The remainder of this discussion principally focuses on T&E required to support the FAA's regulatory process, although much of the approach applies to developmental testing of engineering prototypes as well.

As required by Public law 101-604, Section 107, the FAA reviews the threats to civil aviation with particular focus on explosive materials which present the most severe threat. This review results in a list of the minimum amounts, configurations, and types of explosive materials which would reasonably be expected to cause catastrophic damage to commercial aircraft. In its regulatory role, the FAA must prepare and issue guidelines for use of explosive detection devices and requirements for explosive detection systems to the U.S. air carriers. The FAA must issue guidelines relating to the performance characteristics of prospective devices (EDD) that would allow them to be configured in different system architectures. Regulations should define specific operational requirements for systems, including probability of detection for those types and quantities of explosives that comprise the terrorist threat. The testing program must support both system and device evaluation, validation, and qualification for certification.

An important distinction is made regarding the categorization of the equipment being tested. An Explosive Detection Device (EDD), which employs a particular instrumental method to detect the presence of explosive material, would be tested by the FAA for qualification/verification purposes. The objective of this testing is to verify its performance as matching the data provided by the manufacturer or, in some instances, simply determine its performance. The result would be a set of parameters that characterize the operational performance of the device. An example of parametric testing is the generation of a family of curves relating probability of detection to false alarm rate for different sensitivity settings of the equipment.

On the other hand, an explosive detection system (EDS), which could be comprised of one or more integrated EDDs, would be tested for certification. The objective of this testing is to certify that these instruments, together with their underlying system architectures, meet the FAA-required EDS performance standards under realistic airport operating conditions. The actual testing would be conducted similar to parametric testing, but rather than generating a family of curves, only the test conditions mentioned in the specification would be measured.[1] The test results will be pass/fail.

Only certified EDD equipment would be available for installation at airports. It is probable that EDSs will be offered by systems integrators who would "package" instruments produced by others, together with the necessary baggage transport mechanisms, computer hardware and software, etc. The "package" as a whole would be certified; whoever submits the EDS for testing will hold the certificate.

FAA's requirements for an EDS certification program should encompass more than certification testing of an initial EDS. The certification program should include requirements that the EDS vendor address key aspects required to provide assurance that each EDS unit sold would perform over a period of time in the field at least as well as the one that passed the certification test. Additional certification factors would include: accountability for manufacturing quality control; configuration control of hardware and software; calibration and procedural data; maintenance requirements; personnel training;[2] etc. The certification process should allow for incorporation of operational improvements into devices and systems, as well as cost and size reductions. The certification program must encourage participation by air carriers, airport operators, and equipment vendors, along with the FAA, in the rule making process. A comprehensive certification program is strongly recommended by the committee.

Establishment of the explosive detection system (EDS) performance standard[3] should include a limited number of minimum technical performance specifications for the following parameters: detection probability; false alarm rate; and throughput rate. Size, weight, support requirements and other important operational parameters should be specified by an allowable range of values. This range can be narrowed for specific airport sites. Once the FAA is certain that at least one EDS can meet the EDS performance standard (having more than one available EDS is highly desirable), a schedule of required implementation dates by airport location can be established. The EDS standard developed by the FAA should also describe countermeasures that terrorists are likely to employ once the general nature of the EDS equipment is known.

The FAA will be responsible for routine and random testing of operating systems in the field to ensure that deployed systems are being properly maintained and are operating in compliance with established standards. The section on the testing organization discussed below does not address the enforcement area.

---

[1]Parametric testing could be performed on EDS equipment to fully characterize the operational characteristics prior to the formal certification process.

[2]Human factors considerations must be an integral part of the overall security program. See Technology Against Terrorism--Structuring Security, OTA-511, Chapter 5, "Human Factors in Aviation Security."

[3]Proposed criteria for Certification of Explosive Detection Systems was published in the Federal Register, Volume 57, Number 214, Notices 52698-527002, Nov. 4, 1992. It was assigned Notice Number 92-16, Docket Number 27026.

C. TESTING ORGANIZATION

The future market potential for explosive detection devices and systems in commercial aviation is quite large. Also, the FAA R&D program has actively supported the development of certain instrumental methods. For these, and other reasons, one can readily envision pressures being exerted on participants involved in the FAA certification process. There are at least two ways in which the FAA can minimize the potential influence of these pressures.

- The first approach, and the one recommended by the committee, would require that an independent, well respected organization, outside of the FAA conduct all certification testing. This outside organization would report directly to the FAA Administrator. Candidate organizations include federal laboratories with current testing responsibilities for other federal entities, and independent research institutions with existing testing expertise. This outside organization must not be directly involved in the development or manufacture of explosive detection devices.

- The second approach involves an organization within the FAA that would conduct the testing. This organization must have established expertise in testing matters using specified protocols and sound scientific principles. Because of the importance and impact of testing, it is essential that it be independent of, and insulated from, pressures, whether real or perceived, arising from sources internal or external to the FAA. For example, the FAA must not appear to give preference to the testing of an EDS which employs a technology developed as part of an FAA-sponsored R&D effort.

The activity of this independent testing organization must be overseen by an advisory board with expertise in technology and testing whose composition is independent of the FAA. This advisory board would report directly to the FAA Administrator. For each test, the board would review the test plans, the test data and the test findings and interpretations, and attest to the accuracy of the report's findings and conclusions.

D. TEST PLANNING AND PREPARATION

The important first step in planning any test is the definition of requirements. These requirements must be based on key technical performance indicators derived from the EDS performance specification. These indicators should encompass the most significant elements contributing to the performance of the equipment and be directly measurable.[4]

Activities involved in test planning should include:[5]

- Definition and schedule of all test requirements.

---

[4] *Systems Engineering Management Guide*, Defense Systems Management College, Dec 1986, pages 14-1 to 14-19.

[5] B. S. Blanchard, *Logistics Engineering and Management*, Prentice-Hall, 1986, pages 238-267.

- Definition of the test management approach.

- Definition of test conditions and logistic resource requirements; this includes the test environment, facilities, support equipment, test personnel, test procedures, etc.

- Description of the test preparation phase; this includes test method, training of test personnel, preparation of test facilities, etc.

- Description of the formal test phase; this includes test procedures, test data collection and analysis.

- Test documentation.

- Test funding requirements.

Rather than every test individually addressing each of the above areas, it is usually much more efficient if the common elements that address the technical aspects of a test are abstracted into a general testing protocol. A general protocol insures that all tests consider the same critical factors in a consistent way. The test director, together with a team of experts, would use the framework of the general protocol to prepare for a particular test, including the detailed test specification. This results in a uniform, fair testing approach.

The committee recommends that standard test protocols and procedures be prepared for:

- Test and Evaluation (T&E) required for the verification of the functional characteristics of an explosive detection device (EDD) or system (EDS). Analysis of the types and causes of false alarms should be included.

- T&E required for the certification testing of an EDS.

- Field level verification/validation testing of EDSs installed at airports to assure that deployed systems continue to meet the standards set for certification.

The committee recommends that for certification testing, as well as for field level verification/validation testing conducted at various airport facilities, actual explosives be used in the tests as opposed to simulants until simulant technology further advances. The use of simulants for bulk detection systems is technologically achievable, but the simulants must be developed and validated.

The remaining discussion in this chapter assumes the following scenario for all categories of testing:

- The device or system presented to the FAA for testing would first have been tested by the manufacturer using, to the extent possible, the same protocols and performance criteria that the FAA will use. These procedures and results would be reviewed by the FAA before a candidate device or system would be accepted for test. The device or system would have been subjected to an analysis regarding susceptibility to likely countermeasures, and a determination made that the device or system could not be readily defeated.

- An independent test director with assistance from the test team and the FAA would prepare a detailed test procedure. This procedure would meet all the requirements of the appropriate General Test Protocol and, as applicable, the FAA Certification Program.

- The independent test director would be responsible for conducting the test, making and documenting any deviations from the test procedure that are necessary during the test, and preparing the final report.

- Distribution of the test results would be the responsibility of the FAA, but the results would be available to the vendor.

As mentioned previously, testing can take the form of pass/fail determinations in which the vendor's device performance specifications are measured against fixed criteria. Testing can also involve parametric measurements in which characteristics of importance are verified at specified levels of statistical confidence. It is important that the test team retain all EDD test data so that any potentially relevant probabilities can be estimated. In this manner, parametric measurements of EDDs would allow a system architecture to be analyzed with several different configurations and a system functionality to be calculated.

The FAA asked the committee to provide a general test protocol that could be used for evaluating instrumental equipment that detect explosives via bulk properties[6] as well as vapor properties. However, a general protocol for vapor devices was not possible.

In contrast to bulk detection, the presence of an explosive is only inferred by vapor detection. From vapor detection alone, nothing can be said about the amount of the explosive present. For instance, vapor in sufficient quantity to cause the current generation of detectors to alarm can come from crumbs of explosive material. Conversely, amounts of explosive ten times the lethal threat quantity, if properly encapsulated, would not necessarily cause these detectors to alarm because there would be no vapor for detection. Aside from the special cases of complete containment and the availability of particulate material, the amount of vapor available for sensing from a given quantity of a concealed explosive is not known. At this time, there is no certified vapor generator available to produce a known small quantity of an explosive's vapor, and there is no reference instrument available to check these generators or the background contamination at low levels.

The committee began work on a general protocol for verification and certification testing of bulk EDS systems. An acceptable general protocol must specify the main factors and considerations required to develop detailed test procedures, conduct the test, analyze test data, and evaluate the tested

---

[6]In this context, bulk detection refers to the sensing of some physical or chemical property of the object under examination.

equipment. The protocol must be general to accommodate the testing of the various instrumental devices and systems yet it must encompass all the important factors to be considered so that the use of the protocol will ensure that all equipment will be tested in a consistent and fair manner.

For testing of explosive detection equipment, assuming an independent test team following accepted test protocols employing a standard threat, standard bag set, and well characterized test facility, the key performance indicators (sometimes referred to as Critical Operational Issues) are:

- Probability of detection.
- False alarm rate.
- Bag processing rate.

The following two sections present the results of the committee's efforts on a general test protocol for bulk detection equipment, and a test procedure for vapor devices. The role of the general testing protocol in the context of the certification testing is shown in Figure 2.1.

E. A GENERAL TESTING PROTOCOL FOR BULK EXPLOSIVE DETECTION SYSTEMS

A draft of a protocol for bulk EDS had been prepared for the FAA Technical Center by a group of independent consultants in 1991.[7] Committee members informally reviewed this draft using the criteria discussed in the previous section as a guideline. Extensive discussions and individual written critiques resulted in changes that were reflected in the consultants' final report to the FAA.

In late 1991, the report was distributed by the FAA to industrial contractors involved in the development and/or production of explosive detection equipment with the comment that this protocol should be used as a guide in their testing efforts since a version of it would be used to develop the FAA's test plans. This protocol requires the formation of an independent test team, clear definition of the performance to be verified, and the creation of a test plan for each system or device. Statistical considerations are discussed. In addition, a standard set of bags and threat types, and the creation of a dedicated test site are recommended.

The "General Testing Protocol for Bulk Detection Systems" presented in Appendix A, has been developed in consultation with the committee. It is a guide in designing detailed verification and certification test and data analysis plans.

The general testing protocol in Appendix A differs somewhat from that distributed by the FAA. Within the protocol, the committee has differentiated between those aspects that pertain to certification testing and those that pertain to verification testing, added to the discussion regarding the use and composition of a standard set of baggage for certification tests, expanded on the issues involved in selecting the test location, included provisions to

---

[7]Dr. Joseph A. Navarro; Mr. Donald A. Becker; Dr. Bernard T. Kenna; and Dr. Carl F. Kossack.

FIGURE 2.1. EDS Certification Testing

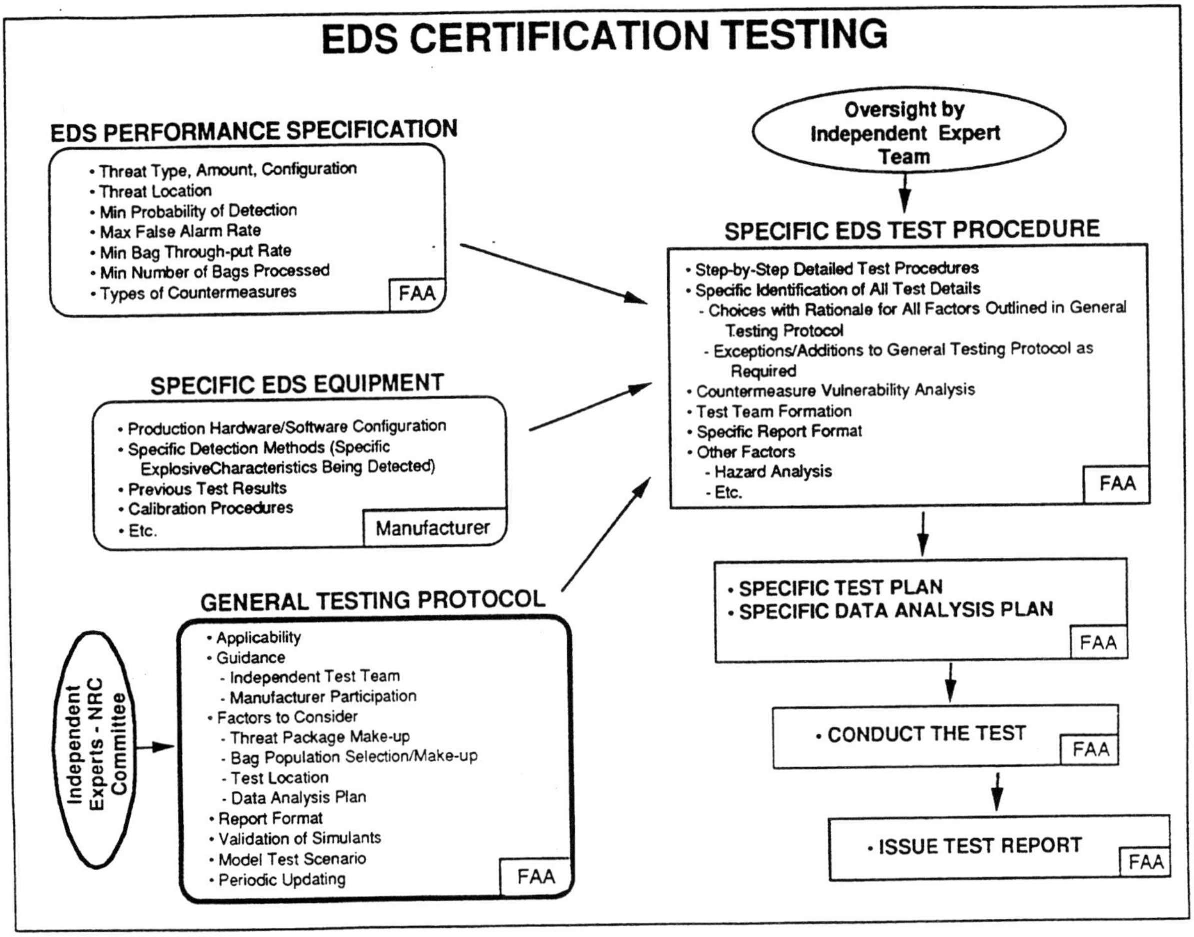

revise the protocol as additional testing experience is gained, added some discussion on testing with countermeasures, required analysis of false alarm data for verification testing, and required the documentation of deviations from the protocol.

The protocol allows for the results to be reported as pass/fail as well as statistically based parametric data. A synopsis of the protocol's primary test conditions is shown in Table 2.1. This general protocol is to be used to plan specific procedures to test the performance of an automatic explosive detection system or device in finding bulk explosives concealed in hand carried or checked baggage. This protocol should not be used for testing instruments based on vapor or particulate detection. The protocol is specific to testing of production hardware, as opposed to proof of principle or developmental testing of early prototypes.

TABLE 2.1 Certification Versus Verification Operational Testing

	Test Outcome	Type Of Equipment	Test Location	Threat Package	Bag Population	Test Time
EDS Certification Testing	Pass/ Fail	Low rate or full-scale production units	FAA Dedicated Site	Live Explosives, types and quantities specified in the FAA's EDS Requirements Specification	FAA Standard Set	Limited Duration
EDD Performance Verification Testing	Parametric Data on Functional Characteristics	Low rate or full-scale production units	FAA Dedicated Site, or Airport Environment	Live Explosives, or Simulants (at Airport Sites)	FAA Standard Set, or Actual Passenger Bags	Limited Duration, or Extended Duration (at Airport Site)

The protocol identifies and addresses the significant factors that must be considered in:

- Designing a test plan that provides data to allow for a fair and consistent verification of the operational functional characteristics of the tested detection devices, or certification of a bulk detection system. Specifically, the protocol identifies the factors that can influence the actual conduct of the test and bias the measurements, thus affecting the results. At a minimum, these factors include: the identification of the characteristics of the explosive material that is being measured by the equipment; the identification of the threat explosives and the representative bag population that will be used for the test; and selection of the test location, equipment calibration plan, and description of the roles of the various participants before, during and after the conduct of the test.

Developing the data analyses plan should always be completed prior to the conduct of the test. Key factors include: planning for the data collection and recording; selection procedures for specific test bags (composition and quantity); and identification of the specific threats that will be tested in which bags.

Conducting the tests. Areas addressed in the protocol include: description of how the tests should be conducted to ensure that the equipment manufacturer can not affect the test conduct or results; procedures to test for reproducibility of results during the test runs; and specific details of the test team roles/duties during the test period.

Analyzing the test data and preparing the report.

The committee recommends that this protocol be reviewed annually for updating at appropriate intervals by outside experts to incorporate knowledge gained as more testing is accomplished so that it becomes "a living document." Over the course of a year, draft versions of the protocol assisted in the planning of verification tests, and that experience was very useful in pointing out areas in the protocol that needed revision. This process should be continued in the future.

F. THE STATUS OF VAPOR AND PARTICLE DETECTION TEST PROTOCOLS

The previous NMAB report,[8] recommended that a clear definition of performance criteria for detector systems be developed, and that standard vapor generators be built to deliver known amounts of vapor to devices under test. This has not been done for vapor systems; and therefore, a general testing protocol cannot yet be developed. The detailed technical understanding of the mechanisms involved in the evolution of and subsequent behavior of vapor leaving an explosive device is lacking. The false alarms caused by unintentional airborne interferrants require that special facilities and additional detectors be available for monitoring during tests. The required protocol for bulk explosive detection (Appendix A) is different than that required for vapor systems and our understanding of what should be tested is too limited to establish a generic test protocol for vapor devices or systems.

The measurement problem is made more difficult by the lack of a standard vapor generator to present a known amount of material to a device for testing. A reference instrument has not been demonstrated that can detect explosive vapor at the lower limits (50 femtograms) to assure the vapor standard performance during tests. The consistent complaint of vapor device manufacturers that background contamination unfairly affects test results can only be settled by the use of such a reference instrument. The reference instrument must have a limit of detection that is lower than that of the devices being tested, and must have a chemical specificity that exceeds that of the candidate systems.

In an attempt to increase the understanding of vapor detection, the FAA is establishing a test facility at Idaho National Engineering Laboratory (INEL). Experience in this facility could provide basic knowledge of vapor detection mechanisms under different scenarios of sample preparation, handling, and concealment. Knowledge of the amount of material available for

---

[8]National Research Council. 1990. Reducing the Risk of Explosives on Commercial Aircraft(U), NMAB-463. National Materials Advisory Board, Washington, DC.

sampling would allow rational assessment of the technology needed. If detectors were successful in detecting this level, a general test protocol could then be written to address the need.

The committee recommends that, if vapor detection systems are to be certified, a test facility must be maintained that has a standard vapor generator and a reference instrument. The facility will have to be free from contamination and capable of remaining in that condition after the tests.

REFERENCES

Ashley, Steven, "Safety in the Sky: Designing Bomb Retardant Baggage Containers," Mechanical Engineering, June 1992, pp. 81-86.

Crossberg, Mitch, "FAA Aviation Security Human Factors," Briefing to Air Transport Association Airline Security Group, FAA Technical Center, Aviation Security R&D Service, Atlantic City, NJ, Sept. 1991.

Gosling, G.D. and Hansen, M.M., "Practicality of Screening International Checked Baggage for U.S. Airlines," Institute of Transportation Studies, University of California at Berkeley, July 1990.

National Research Council, "Reducing the Risk of Explosives on Commercial Aircraft(U)," Committee on Commercial Aviation Security, NMAB-463, Washington, DC, National Academy Press, 1990.

President's Commissions on Aviation Security and Terrorism, Report to the President, Washington, DC, May 15, 1990.

Rechtin, Eberhardt, Systems Architecting: Creating and Building Complex Systems, Prentice Hall, 1991.

Ternes, J.W., and Prestrude, A.M., "Integration of the Human, Canine, Machine Interface for Explosive Detection," Presented at First International Symposium on Explosive Detection, Thursday, November 13-15, 1991.

U.S. Congress, Office of Technology Assessment, Technology Against Terrorism: Structuring Security OTA-511, Washington, DC, U.S. Government Printing Office, January 1992.

U.S. Congress, Office of Technology Assessment, Technology Against Terrorism: The Federal Effort, OTA-481, Washington, DC, U.S. Government Printing Office, July 1991.

APPENDIX A

A GENERAL TESTING PROTOCOL FOR BULK EXPLOSIVE DETECTION SYSTEMS

Developed in Consultation With
The Committee on Commercial Aviation Security
January 1993

CONTENTS

A1

INTRODUCTION

In March 1990, the FAA Technical Center established a task force of independent consultants[1] to undertake testing and evaluation of the Thermal Neutron Analysis (TNA) technology as implemented by SAIC at Kennedy International Airport in New York (JFK). Based on the results of that work, the task force was then asked to prepare a protocol for the conduct of operational testing of explosive detection devices (EDD) or systems (EDS) for checked or carry-on airline baggage (bags, containers, etc). The FAA also intended to use the protocol in the certification of bulk explosive detection systems.

This version of the protocol has been developed in consultation with the National Research Council's Committee on Commercial Aviation Security. The committee has differentiated between those testing aspects that pertain to certification and those that pertain to verification testing; added to the discussion regarding the use and composition of a standard set of baggage for certification testing; clearly recommends a FAA dedicated test site for certification testing; requires that the rationale for deviations from the protocol be documented; and provides for revisions to the protocol as additional testing experience is gained, added some discussion on testing with countermeasures, required analysis of false alarm data for verification testing, and required the documentation of deviations from the protocol.

This protocol is specific to testing of production hardware, as opposed to developmental brass/bread board models or prototype versions of the equipment. This protocol is applicable to:

- o the testing of devices or systems that are automated (i.e., no human intervention used for the detection process);
- o test methods that do not change the characteristics of the item as a result of the test;
- o test methods that detect explosives and explosive devices via bulk properties (e.g., vapor detection devices are excluded).

This protocol does not provide sufficiently detailed plans and procedures to allow the testing and evaluation of a specific hardware device or system. However, the protocol provides the guidelines and framework for planning more detailed test procedures.

---

[1]Dr. Joseph A. Navarro (Chairman); Mr. Donald A. Becker; Dr. Bernard T. Kenna; and Dr. Carl F. Kossack.

As experience is gained in the testing of equipment, this protocol should be updated annually to incorporate additional or modified guidance.

For each application, the FAA will establish the specific threat package (including size, shape, amount and type of explosive) to be detected. Although there will only be one overall threat package, one could envision (in a long range plan) that technologies could be appropriate for, or apply to, a subset of the threat package but not the total package. The FAA will also provide a description of any likely countermeasures that the equipment should be tested against.

This protocol addresses two types of testing: Pass/Fail testing, required for certification testing; and parametric testing, used to obtain statistically-valid verification performance data. Table A1 summarizes the primary differences between these two types of tests.

TABLE A1. Certification Versus Verification Operational Testing

	Test Outcome	Type Of Equipment	Test Location	Threat Package	Bag Population	Test Time
EDS Certification Testing	Pass/ Fail	Low rate or full-scale production units	FAA Dedicated Site	Live Explosives, types and quantities specified in the FAA's EDS Requirements Specification	FAA Standard Set	Limited Duration
EDD Performance Verification Testing	Parametric Data on Functional Characteristics	Low rate or full-scale production units	FAA Dedicated Site, or Airport Environment	Live Explosives, or Simulants (at Airport Sites)	FAA Standard Set, or Actual Passenger Bags	Limited Duration, or Extended Duration (at Airport Site)

In order for the FAA to make a decision on the operational functional characteristics of the device for systems, the FAA must consider:

o estimated probability of detecting explosives $p(d)$, (as observed in the testing of the EDD/EDS.
o estimated probability of false alarm, $p(fa)$, (as observed in the testing of the EDD/EDS).
o estimated processing rate of the bags, r.

The FAA may also be interested in the trade-off between the two probabilities, $p(d)$ and $p(fa)$, especially for those approaches that can readily adjust detection thresholds (thus affecting these fractions).

Other factors that should be considered by the FAA, in determining the characteristics of the detection equipment, include:

o reliability/maintainability/availability,
o cost, initial and recurring; size; and weight,
o significant operational constraints (environment, manpower, etc.),
o bag processing time distribution,
o as appropriate, false alarm types and causes.

The FAA will: provide the test team (which will be responsible for generating the specific test plan preparing and executing the test, analyzing and evaluating the test data, and preparing the report on the findings of the test), and establishing where and when the test will take place.

The test team should have a test director and be composed of experts in the technology being tested, test and evaluation planners, and analysts who can design the statistical plan and conduct the evaluation of the test results. An independent observer should also be a member of the test team. This observer should comment on all activities associated with the testing and evaluation of the EDD/EDS, including: adherence to the test plan, adequacy of the plan, perceived testing limitations, and potential sources of test bias.

All test baggage and test articles, the threat package (explosives or simulants), and personnel will be provided by the FAA.

Chapter A2 of this protocol provides some general requirements associated with the operational testing process. Chapter A3 addresses a set of issues that must be considered and specific requirements that must be fulfilled prior to the development of a detailed test and evaluation plan. In Chapter A4, specific aspects of the detailed plan are discussed. Chapter A5 deals with issues related to the conduct of the test, while Chapter A6 discusses the data analyses and evaluations of the test data. Finally, Annex I discusses how the FAA could establish a standard set of bags and explosives to conduct operational tests or certification tests at a dedicated FAA test site. Annex II contains a suggested approach for validating explosive simulants and Annex III contains a model testing scenario.

A2

GENERAL REQUIREMENTS

In order to develop a specific test plan, the test team must consider all of the factors that may: influence the conduct of the test; bias the measurements related to the detection equipment under test; and affect the results obtained from the test and/or the reliable interpretation of those results. The test team must establish the conditions under which the test is to be conducted and the characteristics (i.e., attributes or variables) of any bag as it is processed.

There are a number of steps or topics that must be considered before development of a final specific test plan for the equipment under test. These may be separated into two groups: general topics and specific topics. The general topics are discussed in this chapter, and the specific topics are covered in the next chapter. However, since this is a generic protocol for a variety of explosive detection devices or systems (using different technologies), there may be some additional factors that may need to be considered, and the discussion that follows should not preclude additional factors from being included in the final test plan if those factors are considered to be relevant by the test team.

It is assumed that prior to this point, the following activities have already occurred:

o The specific device or system presented to the FAA for testing would first have been tested by the manufacturer using, to the extent possible, the same protocols and performance criteria that the FAA will use. These procedures and results would then, in most cases, have been reviewed by the FAA for adequacy before the candidate device or system would be accepted for testing.

o The FAA would have appointed an independent test director and a test team to prepare a detailed test procedure from this general test protocol.

o The test team would be expected to conduct the test in accordance with the detailed test procedure.

o An agreement would be finalized with the equipment supplier regarding what results will be provided by the FAA after the testing is completed.

A. Explosive Characteristics to be Measured

After the specific equipment to be tested has been identified, the set of characteristics of the explosives that the equipment will measure for detection must be identified and specified. The physical principles employed by the instrument for detection will determine which physicial characteristics of the explosive material are of interest. This determination is especially important when the use of explosives—such as at airport facilities—will be prohibited during the test of the equipment, so that the use of a simulant should be considered. For example, one of the primary characteristics of explosives measured by TNA devices is the nitrogen content. The test team must determine if simulants can be identified which will exactly mimic the characteristic nitrogen content for the equipment under test.

In addition, any countermeasure techniques to be included in the testing should be identified prior to the test initiation by the FAA, or by the test team in consultation with the FAA. Some technologies are relatively easy to countermeasure while others may be more difficult.

B. Identification of the Set of Threat Explosives

The set of threat explosives (type, shape, and weight) to be used in the testing must be specified by the FAA. For example, the FAA may require that testing must include x pounds of sheet explosive of RDX/PETN base (Semtex). The FAA must also specify the relative frequency of expected occurrence for each item in the set of threat explosives.

The FAA should identify the placement location for the explosives in the containers. During the course of testing, it may be observed that the instruments response is sensitive to the placement location of the explosive in the bag. If this occurs, the FAA should consider additional testing with the threat in most disadvantageous locations.

C. Identification of Potential Bag Populations

The test data base should contain observations for the test bags of all the major characteristics that will be measured by the detection system being tested. This database will be used by the test team to select representative groups of test bags. The actual set of bags used for testing could be: (1) actual passenger bags; (2) fabricated by the FAA appointed team; or, (3) selected from the set of FAA "lost" bags. Bag selection is a crucial topic in designing a test plan that is fair and effective.

For certification testing of an EDS, the testing should be performed on a standard set of bags to provide a fair and consistent comparison against the EDS Standard. Annex I discusses how a standard set of bags could be established.

For non-certification testing, the goal will be the generation of parametric performance data. The test team must determine the characteristics of bags typical of those that will be processed at a designated airport. If the testing itself will not be conducted at the airport, data on actual passenger bags being processed at those facilities could be collected. If these data were collected over a sufficiently long period of time, the effect of seasonal and bag-destination differences on bag characteristics could be determined.

Additional information on bag selection techniques is contained in Chapter A3.

D. System Calibration and Threshold Settings

This protocol applies to devices or systems that are totally automatic in their response; i.e. operator independent. The manufacturer will not be allowed to change or modify the settings of the equipment once the test for a given bag population has been initiated. Thus prior to the start of the test, the manufacturer should be allowed to have access to the set of bag populations that will be used for the testing so that they can determine the associated response of the equipment to the characteristics, and thus calibrate and establish the threshold settings of the equipment. The manufacturer should provide the FAA with the complete calibration protocol.

The manufacturer should not be provided details regarding the relative frequency of threat occurrence and location of the threats in the bags, since the tests must be as blind as possible.

While the manufacturer is establishing the threshold settings, they should also be required to provide the FAA test team with the *p(fa) versus p(d)* relationship as a function of the threshold setting. This data should be made available prior to the testing unless the equipment can store the basic data so that, after the fact, the *p(fa) versus p(d)* curves for the different threshold settings could be reconstructed.

E. Manufacturer/Contractor Participation

Although the test team may be required to rely heavily on manufacturer or contractor personnel for support in conducting the testing, procedures should be established to minimize the possibility that they could influence the test results. Toward this end, the manufacturer may be required to train FAA chosen personnel to operate the equipment during the test. These personnel should have the same general skill levels as those who will be expected to operate the equipment in the airport environment.

F. Test Sites

For EDS certification testing, an FAA-dedicated test site is required that can be controlled and characterized, particularly with respect to background contamination. Explosives representing the standard threat set must be used. The standard bag population also must be used. Similar considerations apply to EDD verification testing at the FAA test site, although a bag population representative of a specific airport may be used.

For testing conducted at airports, available passenger bag populations could be used. If tests are conducted over an extended period of time, the effect that various seasons have on the characteristics of the bags can be determined. In most cases, the use of explosives will not be acceptable and simulants will be required. (Simulants would also be needed to monitor the performance of the devices or systems while they are operational at airports.)

A major disadvantage of airport testing is that it could interrupt operations in the airport terminal. On the other hand, if the equipment is to be used in conjunction with other equipment already in place at the airport—such as baggage handling equipment or another EDD—time and expense could be saved by not having to provide the other hardware at the off-airport site.

G. A Standard Set Of Bags And Threats

At the present time, the FAA has not developed a standard set of bags, a standard suite of explosives, and a dedicated test site. Annex I suggests how the FAA can establish these three items for EDS qualification testing. This protocol also addresses other approaches so that testing can proceed prior to the establishment of any or all of these test resources.

A3

SPECIFIC REQUIREMENTS

Before a test plan can be developed, the test location and the constraints on the use of threats must be known. For certification testing, the preferred location of the tests should be the FAA dedicated site. Furthermore, it is preferred that actual explosives samples be used when testing for the detection capability of the equipment. Finally, it is preferred that standard bags be used when testing for both the detection and false alarm capability of the equipment. Unfortunately however, it may not be possible to conduct the tests in the above preferred manner. The test team must first determine what deviations will take place and use alternative methods for achieving as realistic and meaningful tests as possible, given these deviations. The test director will identify the deviations and provide supporting rationale.

If the tests are conducted at FAA test facilities, it most likely will not be possible to use passenger bags to estimate the fraction of false alarms, P_{FA}. It may also be difficult to fully address operational processing rates of bags. If, on the other hand, the tests are conducted at airport facilities, it may not be possible to use actual explosive samples, and the use of simulants may be required. Under either situation, the test team must determine the set of bags that should be used in estimating both P_D and P_{FA}. The selection process will depend on the purpose and location of the test.

The following three sections discuss specific areas that must be addressed before the test team can develop the detailed test plan: identification of the set of distinct bag populations that must be used in the testing of the equipment; selection of the threat package to be used; and, specification of the procedures used to measure baggage processing rates. In addition, a short discussion of the issues involved with pre-testing is presented.

A. Identification of Distinct Bag Populations

One of the most important aspects of the test plan and subsequent analyses is the selection of the bags to be used for the tests. These bags must reflect the types that will be operationally encountered and processed by the equipment, over time at various locations. From experience, it is known that bags destined for one location, at a given time of year, are packed with different items than bags going to another location, or even going to the same location during a different season. Also, tourists tend to pack different items than business travelers. The contents of the processed bags, and the effect that background levels and interferrents can have on the measurements taken of the characteristics of the explosive, must be considered in any test plan. Toward this end, the FAA will specify the constraints/conditions placed on measurements associated with estimating P_D and P_{FA}. For example, the size and weight of a test bag may be restricted, their destination and/or time of year may be specified. Depending on the device, other constraints/conditions will have to be determined.

The test team must address the issue of the number of different bag populations that the detection equipment will process during the test. For testing which does not involve EDS certification, different applications or situations may be such that the performance of the equipment will change significantly from one application to another. If it is determined that these differences are important to the estimation of P_D and P_{FA}, then the different bag populations should be used for the OT&E testing.

After the test team has identified the potential set of bag populations and established the various factors that are significant in determining the detection of an explosive, the team must collect data to finalize the test bag set. In order to evaluate these bags, the data collected should include the numerical values of the measured characteristics of the bags.[2] For example, the TNA device measures among other things, the quantity and distribution of nitrogen in the passenger bag.

In order to collect parametric data, the equipment under consideration can be physically located at an airport facilities for which the potential baggage populations can be observed or at the FAA dedicated site using a standard bag set. For each of these locations, the equipment manufacturer should be informed by the FAA what maximum fraction of false alarms will be acceptable, so that the equipment can be properly calibrated and the detection threshold established.

For testing at an airport, the test team will determine who will process passenger bags from these populations to provide the observed measurements of the explosive characteristics for each processed passenger bag. A sufficiently large sample of processed bags is required, so that a reasonable estimate of the multi-variate frequency distribution of the set of characteristics being measured (for each of the populations) can be obtained. At the same time, data is being collected for estimating the P_{FA}, which could be of use to the manufacturer in establishing and/or reconfirming the calibration and threshold setting. If the distributions for one or more of the potential populations are not statistically different, then the data from those populations should be pooled to represent one population. In this manner, a new set of populations will be established that will represent the final set of populations to be tested. The pooling process cannot now be rigorously defined and will require a considerable amount of judgment.[3] The

---

[2]Refer to Section A2.C.

[3]Refer to discussion in Section A4.C.

rationale used for pooling should be documented by the test director so that an experience base can be developed.[4]

When testing at the FAA dedicated facilities, the test team must prepare a set of bags from each of the potential bag populations. This requires that the test team know the bag characteristics, as measured by the equipment, for each of the populations. To this extent, some uncertainty will be introduced into the testing process, since the true bag populations are defined by the actual passenger bag populations and the test team is generating artificial populations. In this situation, the test team must attempt to prepare these sets of bags to be as close to the populations of interest as possible. Annex I addresses the use of an FAA standard bag set for use at the FAA dedicated test site.

B. Identification and Selection of the Threats

The threat as specified by the FAA for the EDD/EDS should include:

- type of explosive (C-4, PETN, etc.),
- minimum quantity (mass),
- shape (bulk, sheet, thickness, etc.),
- relative frequency of use of each threat,
- the location of the threat in the container,
- the set of containers to be used (bags, electronic devices, etc.) and other pertinent features, and
- potential countermeasures.

As previously stated, the detection equipment must measure a set of characteristics associated with the bag as it is processed and, based only on these measurements, decide if a threat is present. In order to develop a meaningful test plan, the test planners must know the characteristics of the explosive that are being measured and how these measurements might be affected by the container in which the threat is placed. In order to select the appropriate groups of test bags (when this is required) and/or produce simulants of explosives (when this is required) the following information is necessary:

- The characteristics of the explosive being measured.
- Bag-related items that can affect measured explosive characteristics.
- The relationship between the measured values of the explosive characteristics and the shape, weight and type of explosive being considered.
- The distribution of the observed measurements on the characteristics of the explosive and if the different characteristics are related or independent of each other.
- The discriminate function being used to assimilate the observed values of the characteristics into a detect/no-detect decision. This

[4] At some future time, this information can be used to develop a set of general guidelines for bag pooling.

information is particularly crucial if simulants will be required in conducting the OT&E.

If particular equipment measurements are affected by interferrents, it is important to know: if the explosive characteristic is masked (i.e. the explosive characteristic is hidden in the bag); or, if the response is additive (i.e. the observed value of the characteristic of an explosive in a bag is statistically equal to the observed value of the bag plus the observed value of the explosive without the bag).

If possible, testing should be done using samples of the standard explosive threats defined by the FAA. All explosive samples must be verified relative to type, purity, weight, and chemical composition, if that characteristic is significant for the EDD/EDS being tested. Samples for the analyses should be taken from the explosives under the supervision of the test team. The number of samples of each threat type is determined by the test design; i.e., the number of different bags being processed and the variability of the measurements of the characteristics of different samples of each threat type.

If simulants must be used instead of explosives then it is important to ensure that the simulants faithfully represent the explosives relative to the characteristics of the explosives that are measured by the equipment and used for the detection process. In order to do this, each simulant, representing a specific mass of a particular explosive, must be compared to the actual mass of real explosive material with the equipment being tested. The simulant and explosive should both be in the same geometrical shape; i.e., sheet or block.

Once the explosives have been checked and the simulants produced (the number of each simulant type should be determined by the test team and should be adequate to ensure that a sufficient quantity will be available for the testing program), a validation test is required to verify that the simulants faithfully representing the explosive. Obviously, this validation testing cannot be done at a airport facility. Annex II outlines the recommended procedure for the validation testing of the simulants.

C. Bag Processing Rates

A very significant aspect of the operational suitability of a detection device or system is the average time required to process each bag, R. If this time is excessive, the airlines will have difficulty in incorporating the device or system so as to not affect its schedule of activities. While conducting the various tests associated with the estimation of the detection and false alarm rates, it will be necessary to collect data on this operationally important issue.

The first opportunity to collect operational data is when the test team is obtaining data to select the final set of bag populations. During these tests, procedures should be established to ensure that meaningful temporal data is collected. Individual processing times should be collected for each bag, as opposed to the total time required to process a set of bags, so that the mean and the variance of the processing time can be determined.

The second time that operational data can be collected is during the operational tests. It may be more difficult to collect data at this time since there will be on-going test activity (such as: placement of explosives in bags, marking bags, etc.) which might add to the processing times. To avoid these problems, the processing rate data should be generated with that testing associated with the false alarm rate estimation since almost all of the operational processing will take place with passenger bags (not containing explosives). Again, processing time should be associated with each bag, including the re-processing of a bag when an initial false alarm has occurred.

If the above testing is conducted at airport facilities, the processing time data should be fairly descriptive of the operational processing time of baggage. If, however, the testing does not take place at an airport facility then the test team must set up the testing facility to mimic the airport facilities of interest. Here again, some artificiality will necessarily be built into the collection of this data.

The test team should record any malfunctioning of the equipment, unusual processing activity, manual interference with the automated processing function, etc. This information should be reported in the final report on operational testing activities.

D. Pre-Testing

Prior to full scale operational testing of a device or a system, the test team should conduct a pre-test to determine if the device or system can meet the FAA requirements against a standard, but limited, number of target bags randomly intermingled with a standard, but limited number of normal bags. In this manner, the test team can pre-screen equipment without having to proceed to complex operational testing since most test results do not tend to be "near misses" and should easily be sorted out in such a pre-test. The above standards could also serve as a cross comparison set.

A4

DEVELOPMENT OF TEST AND EVALUATION PLANS

A. General Factors

The test team should be aware of all factors that may influence the conduct of the test. They must also understand how the measurements relate to the detection process as implemented by the instrumental technology. At this point, the team should be ready to complete the T&E plan.

The following items address various factors that should be considered in developing the detailed test plan.

- The test design should be robust so that studies of the key operational factors can be made. All planned pre- and post-test activities, as well as the test activities, should be explicitly identified to all concerned parties -- who, when, how, where, why, etc.

- Data collection should be automated (if possible) and also manually recorded by independent data collectors. As a minimum, the collectors should identify and record the order and bag number of each bag entering the detection equipment and record each alarm response and other appropriate data. All data runs should be numbered, and the time of each run recorded.

- The test team should consider the use of barcode labeling techniques, with all reading done by hand-held barcode readers, and the data automatically fed into a central processing unit. All bags, threats, and countermeasure items should be barcoded.

- Test conditions should be clearly understood and agreed to by the equipment operators. Specific evaluation plans, including all test conditions, should be reviewed by the equipment manufacturer and the operators. The FAA test director will determine if the plan is complete and suitable.

- As appropriate, the acceptable minimum detection probability and maximum false alarm probability required to certify an EDS should be clearly identified to the test team. When confidence statements are made, they will be made at the 95 percent level.

- For qualification testing, the types and causes (if known) of false alarms will be recorded. There are generally two types of false alarms: nuisance alarms caused by the detection instrument being "fooled" by a substance that triggers a response in the detection system that is similar enough to that of an explosive material that the response cannot be adequately discriminated from a real explosive; and, false alarms caused by low signal-to-noise ratio when the sensitivity setting is at the maximum (i.e., signal strength cannot be increased any further). Low signal can be caused by: the inherent low response of the detection method for the quantity of explosive material of interest; attenuation of the signal due to a design inadequacy or component failure, etc. High noise can be caused by large background signal inherent in the detection method, by improper design, failure of the electronic system, etc.

- For certification testing, the primary response variable at a given setting is binary, i.e., detection/no detection.

- The procedure to screen bags should be described in detail, including the specific role of the operators during the conduct of the test.

- Procedures should be in place to accurately record the average time required to process a bag. Sufficient data should be collected regarding the time history (e.g. date/time) associated with each tested bag so that the test run could later be reproduced if desired.

- For parametric tests, the relationship between the probability of detection and the probability of false alarms, as a function of the detection threshold level is an important result. Determining this relationship will require the testing of various bag populations at different detection thresholds. The means to adjust the detection threshold of the equipment must be available to the operators, and all such adjustments recorded.

B. Data Analysis Plan

The specific analysis plan should be developed prior to testing. The analysis plan must describe the data that will be required to be collected during the test, show how the data will be analyzed, and describe the statistical tests that will be used in analyzing the collected data. It is recommended that the data analysis plan be validated by generating artificial data and performing the analysis on that data.

The resulting data collection and analyses plan should be a principle part of the test plan. The test team should be allowed flexibility in case of unanticipated data outcomes, and thus should not be rigorously held to the initial plan. However, each deviation from the plan should be documented with supporting rationale. In any event, the initial hypotheses and the associated criteria must not be changed.

C. Selection of Test Bags

In order to determine the quantity and the characteristics of the specific bags to be used in the testing, a statistically valid approach must be followed. The final set of bags will be selected by the test team from a distinct population of bags (generated by the process described in Chapter A3) for which the data collected on bag characteristics will have been collated to statistically describe the multi-variate frequency distribution of the characteristics.

The set of bags to be used could be selected from FAA-held "lost" bags, fabricated by an FAA appointed team, or actual passenger bags. When non-passenger bags must be used, it would be beneficial if the test team could use several groups of bags for each given population, each group representing a typical set of bags for that population.

One of the following approaches should be used for selecting the final bag set.

- Select Representative Bags When Non-Passenger Bags Are Required. For each of the different populations, designated in Chapter A3, a set of test bags must be available that reflects the multi-variate frequency distribution of the characteristic measurements of bags from that population. A stratified sample of bags from that population should be used; i.e. for each characteristic, its range is partitioned so that an equal frequency of observations are observed in each of the partitions. This generates multi-dimensional cells such that the marginal frequencies of occurrence of each characteristic are equal. Then the multi-variate frequency in each cell is observed and a proportional number of bags are selected to represent each of the cells. For example, if 20 cells are established (each cell representing 5 percent of the population which has the multi-variate frequency of the characteristics as represented by that cell), one bag should be selected from each of the cells, each bag having the characteristics associated with that cell. This would generate a sample of 20 test bags representative of the population. If a sample size of 40 is required, then two bags should be selected from each cell, etc.

  If it is not possible to do the above—for example because the number of characteristics is too large to allow for this approach—drop the least critical characteristic and continue, reducing the set of characteristics until the process can be accomplished. The unaccounted-for characteristics can then be handled by statistical techniques. Alternatively, ensure that at least one bag is selected for each of the non-zero cells, and then use the same statistical techniques to account for the non-representative sample of bags.

  Another possibility is to estimate the correlation between the variables being measured and attempt to use a transformation function, which might assist in the above process.

- Select Bags At Random. If the above process is not possible because of the large number of characteristics or not having an understanding of the correlation between the characteristics, etc., statistically random samples can be taken from the population of bags.

- Use Actual Passenger Bags. If the equipment is set up at an airport terminal (handling bags from one of the chosen populations) passenger-checked bags can be used to estimate the false alarm rate (as discussed in Chapter A3). To test for the detection rate, it may sometimes be possible, with the cooperation of the passengers, to use their bags for incorporating explosives (or simulants).[5]

Alternatively, a "Red Team" can be used to package bags with explosives or simulants that appear to be headed for the same destinations as the actual passenger bags. These bags should be clearly identifiable to the test team (but not the equipment operators) so that they can be retrieved once they have been processed. Under this approach, the "Red Team" determines what will be placed in the bag and where it will be located. This approach will provide estimates of the false alarm probability for the time period being observed, using actual passenger bags, and estimates of the detection probability either using the modified passenger bags or fabricated bags.

D. Number of Test Bags Required

The FAA may specify the minimum number of test bags and observations required to be processed for certification testing. In any event, the number of observations required for the test can be computed from the values of the FAA-provided minimum P_D and maximum P_{FA}, the particular statistical evaluation methodology employed, and the associated desired levels of statistical confidence.[6]

Confidence interval estimation is a standard method used in classical statistical inference. (In general, an interval estimate of a population parameter is a statement of two values between which there is a specified degree of confidence that the parameter lies.) It is normally desirable to have a narrow interval with a high confidence level. Using this approach to calculate the number of observations required to estimate P_D, assuming a 95 percent confidence level with a confidence interval half-width of 0.03, the required sample size of bags with explosives can be approximated by the relation:[7]

$$\left[\frac{1.96 * P_D * (1-P_D)}{0.03}\right]^2$$

Substituting for the constants and ignoring the higher order term, the expression reduces to: $\underline{4,300*(P_D)(1-P_D)}$. For a P_D of 0.90, the required sample size would be 387 bags containing explosives.

---

[5]This can only be accomplished if the threat can be realistically utilized on the outside of the bag to measure P_D. If the threat must be placed inside the bag this approach should not be considered.

[6]For example, see Snedecor and Cochran, *Statistical Methods,* Iowa State University Press, 1978, Chapters 2, 4, and 8; and Brush, "How to Choose the Proper Sample Size," Volume 12 in the *ASQC Basic References in Quality Control: Statistical Techniques*, Cornell and Shapiro editors, American Society for Quality Control, Milwaukee, Wisconsin, 1988.

[7]Snedecor and Cochran, op cit, pp 58-59, p. 113, and pp 211-212.

Another classical statistical inference method is hypothesis testing. A null hypothesis is a statement about a population parameter; the alternative hypothesis can be a statement that the null hypothesis is invalid. The hypothesis testing methodology chooses between competing choices at a specified significance level. There are two possible sources of error: reject the null hypothesis when it is in fact true (Type I error), and accept the null hypothesis when it is in fact false (Type II error). While the interval estimation method only allows for the control of the Type I error (a 95 percent confidence level sets this error at 5 percent), the hypothesis testing method allows for the control of both Type I and Type II errors. For example, assume that the *null hypothesis* is: "the probability of detection, P_D, of the tested device is equal to or greater than 90 percent." The *alternative hypothesis* can be formulated as: "the probability of detection, P_D, of the tested device is less than 90 percent." For a P_D of 0.90, setting the Type I error at 5 percent, setting the Type II error at 5 percent and evaluating it at P_D of 85 percent, the sample size required to discriminate between the two hypotheses is estimated to be 470 bags containing explosives.[8]

The above methods can also be used to determine the number of non-explosive containing bags required to estimate the false alarm rate. The specified minimum P_{FA} should be substituted for P_D, and the same formula applied.

When non-passenger bags are being used, it is recommended that the number of observations be obtained by sampling a given bag without the explosive and the same bag with the explosive threat a minimum of six times each. Thus approximately 1/6 if the above number of calculated bags would be required for the number of observations. It is further recommended that these bags be chosen such that there are multiple groups (i.e., 6 groups of bags) which are each representative samples of the population being considered.

If the "Red Team" approach is being used, it may not be feasible to test each bag with and without explosives.

E. Selection of the Number of Threat Articles

The actual number of threat articles (explosives or simulants) to be used for detection determination will depend, in part, on the number of bags available to conduct the testing. For each defined threat, multiple samples may be required to facilitate the testing. For example, if there are 5 distinct threats, then 4 samples of each threat would be required to provide 20 test articles.

The determination of the number of detection test articles must be made early in the planning process if simulants are required, so that the simulant validation testing can be completed prior to the testing and a suitable amount of the simulants made available.

---

[8]Brush, "How to Choose the Proper Sample Size," op cit, Equation 5.3, page 20.

A5

TEST EXECUTION

At this point, the test team should have accomplished the following:

o Examined and studied the explosive detection equipment to be tested, and fully understand the relevant technical issues and characteristics of the equipment. (If the test team does not fully understand the equipment to be tested, they may not be able to assure the FAA that the equipment has been adequately tested.)

o Examined the raw data and processed data pertaining to the critical characteristics of the explosives and the baggage that will be used in the measurements.

o Selected the set of bags to be used in the testing and understand how these bags represent the population of bags being considered.

o Determined the threat samples and countermeasures to be used during the testing.

The following areas should be considered by the test team in the execution of the testing:

o Independence from the Manufacturer

Although the test team may be required to rely heavily on factory personnel for support in conducting the tests, procedures must be established to minimize the possibility that the contractor could influence the test results. Toward this end, at least one member of the test team should participate in:

- supervising and overseeing all operations of each test including placement of threats in bags;

- numbering and sequencing bags through the tested equipment;

- identifying and manually recording data on each bag as it was being processed;

- verifying that any internal computer identification system does not affect the measurement system; and

- disconnecting computer modems and any other outside manipulative devices to isolate the equipment from the outside world.

As required, the equipment manufacturer should train FAA selected personnel to operate the equipment during the test. The personnel chosen should be representative of types that are expected to operate it when it is in the field, in terms of education, experience, skill level, etc.

o <u>System Reproducibility</u>

During all test periods, the test team should collect data for the purpose of assessing measurement reproducibility by taking repeated measurements of a controlled set of bags. For example, during each test period a set of control bags should be repeatedly processed before, during, and at the conclusion of each days test, in order to provide assurance and documentation that the detection response has not changed significantly during the test sequence. If the test team determines that the response has changed, the test may have to be repeated unless an acceptable answer to why it occurred can be provided to the test team.

o <u>Other Areas</u>

- The test team shall directly supervise the placing, moving or removing the threats in the bags being tested.

- The test team should control the threats with an established chain-of-custody procedure.

- At the end of each test day, print outs of all test data should be collected by the test team; a backup package should be available for the FAA, and this data retained for possible future use/confirmation.

- All available data should be collected, even if it will not be used immediately. At a minimum, threshold values and alarm/no-alarm readings on each item tested should be recorded. All parties should sign off on the data packages each day.

- The equipment manufacturer should be informed of all data requirements (by the test team) as early as possible.

- Prior to any testing, the team should dry run each significant set of test conditions to insure that the equipment is properly functioning in an operational mode.

- If possible and appropriate, all tests and test activities should be video/audio recorded.

- The equipment being tested will be used as set up by the manufacturer; i.e., there will be no non-routine changes or adjustments made during the evaluation testing.

Annex III contains a model test scenario that could be used to guide actual testing.

A6

ANALYSIS AND EVALUATION OF TEST DATA

Once the tests have been completed, the test team will be required to prepare a report of the test findings. It is important that the original test data for detection devices (EDD) be retained so as to allow the FAA to estimate EDS capabilities prior to certification testing. The testing program should have produced most of the following data and information:*

- o Detection characteristics measured.
- o Det of explosives/simulants used.
- o Bag populations used, including justifications.
- o System calibration results.
- o Threshold settings used.
- o Number of bags used and the selection process.
- o Simulant validation data, as appropriate.
- o Detections observed data.
- o False alarm observed data, including for verification testing, the types and causes (if known) of false alarms.
- o Bag processing rates data.
- o Observed sensitivity of instrumental response to placement location of explosives in a bag.
- o Vulnerability to countermeasure as appropriate.

Using well known statistical procedures.[9-10]

- o The fraction of detections observed, p(d) for different bag populations and different threats.
- o The fraction of false alarms observed p(fa) for different bag populations.
- o The estimated processing rate of the baggage, r.

For illustrious purposes, if one assumes that all tested bag populations are equally likely, stratified sampling of the bag populations is used, and all threats and threat locations tested were equally likely, then unbiased estimates of $P_{D,}$ $P_{FA,}$ and r are given by:

*NOTE: Some of these results may be classified in accordance with the FAA's Security Classification Guide.

[9]Snedecor and Cochran, Statistical Methods, Iowa State University Press, Ames, Iowa, 1978.

[10]Hamburg, Statistical Analysis for Decision Making, Harcourt Brace Jovanovich Publishers, 1987.

o p(d)=(# of detections)/(total # of possible detections),

o p(fa)=(# of false alarms)/(total # of possible false alarms),

o r=(time required to process all bags)/(# of bags processed).

If any one of the above assumptions is not true, then these fractions must be computed for appropriate subsets of the test data and these fractions weighted to account for any differences (for example, if the bag populations or threats are not equally likely). At some point the test team should provide a composite fraction for the test for use by the FAA.

Finally, the test team report should contain all of the raw data generated during the course of the testing, as well as the reduced data, and include all of the detailed calculations showing the complete test design and how the results were obtained from the data collected. This report should be sufficiently detailed and transparent so that any competent technically trained individual could completely follow the test and results to its logical conclusions, and be able to duplicate the test program if desired.

ANNEX I

A STANDARD SET OF BAGS AND THREATS FOR TESTING BULK EXPLOSIVE DETECTION DEVICES OR SYSTEMS AT AN FAA DEDICATED TEST SITE

It is recommended that the FAA take the following actions:

O A Standard Bag Set should be established for use by all EDS tests. The Set should be large, say on the order of 1,000. The bags should be selected from FAA-owned "lost" bags, altering bags as necessary to ensure that the Set contains random samples of actual passenger bags with representative contents for different times of the year (seasonal varations) to different destinations (e.g. cold or hot climates) by different types of travelers (e.g. business travelers, tourists, etc.). They should include different types of bag material, varying contents, different shapes, sizes, and weights.

O A Standard Threat Package should be established which would include the specified threat types, at the minimum weights deemed to be important and in the shapes that are of concern. It is recommended that at least five replicates of each specific threat item be produced with the actual explosive, as well as at least five simulants of each threat type.

O Certification testing should be conducted at an FAA facility, using the above Standard Bag Set and the Standard Threat Package. These items, or a duplicate, should be made available to the vendor prior to the actual testing so that the contractor can determine using parametric tests, the thresholds of the system being offered for test. Since the above items must not be tampered with, they must be sealed in a manner to protect their integrity. The vendor should be encouraged to conduct their tests at airport facilities whenever possible.

Four subsets of bags with the Standard Bag Set should be established, each subset containing bags of the same material, i.e., 100 aluminum bags, 500 cloth bags, 50 leather bags and 450 plastic bags, or whatever is deemed to be operationally representative of the proportion of passenger bags. The contents of the bags in each of the subsets should reflect typical seasonal wardrobes for spring, summer and winter.[11] Approximately one-third of the bags in each subset should contain contents reflecting each of the three seasons. For each subset, approximately one-third should be light in weight,

[11]Spring, summer, and winter are not precise terms. They have been determined empirically from TNA testing in which the nitrogen contents of baggage has been observed to undergo distinct shifts.

one-third should be average in weight, and one-third should be heavy in weight. Table A1 is representative of the above allocation. The FAA will need to determine the proportion of bags according to bag material. Once this is done, the allocation of number of bags in each cell can be determined. This suggested break-out of bags should be updated as additional testing experience is gained.

These bags should then be barcoded and well marked so that they can be easily read by a barcode reader as well as by visual means.

TABLE A2. Possible Make Up Of 1000 Standard Bags

TYPE OF BAG		LIGHT-WEIGHT	MEDIUM-WEIGHT	HEAVY-WEIGHT
ALUMINUM	SPRING SUMMER WINTER	11 11 11	11 11 11	11 11 11
CLOTH	SPRING SUMMER WINTER	56 55 56	55 56 55	56 55 56
LEATHER	SPRING SUMMER WINTER	6 5 6	5 6 5	6 5 6
PLASTIC	SPRING SUMMER WINTER	50 50 50	50 50 50	50 50 50

With respect to the Standard Threat Package, the FAA should also determine if the threats need to be inserted into the bag for test purposes. If not, then threat packages should be developed which can be placed on the outside of the bags which are being tested for threat detection. The threat packages should also be barcoded and well marked. If the threats must be placed inside of the bags, then the threat package should be so constructed so that it will not move inside of the bag.

There should be two different kinds of threats, one in which the bag is the bomb (i.e., the explosive is molded along the side(s) of the bag in sheet form), and the other in which the bomb is placed inside of the bag (i.e., it is not in sheet form).

For parametric testing, the tests should be conducted so as to allow estimation of the P_D versus P_{FA} curve for different detection levels. This will require that the threshold of the equipment be changed several times during the test to accommodate, for instance, variations in the bags. One P_D, P_{FA} point will be (0,0) and another will be (1,1). If three more points can be established, then the FAA will have a reasonable estimate of the equipment capability. Figure A2 depicts a possible outcome of the operational test of a TNA unit.

Parametric testing should give the FAA significant insight as to whether or not the system has a reasonable possibility of passing certification testing. If at the end of the testing, the perceived likelihood of passing the certification test is low, the contractor should be told that no certification testing will be conducted until the FAA determines there is a reasonable possibility of success.

With respect to certification testing, the FAA should, at a minimum, specify the relative weights assigned to each of the threat items, as well as the minimum acceptable average P_D (i.e., across all appropriate weighted threats) and maximum acceptable average P_{FA}. The outcome of this test should be pass or fail.

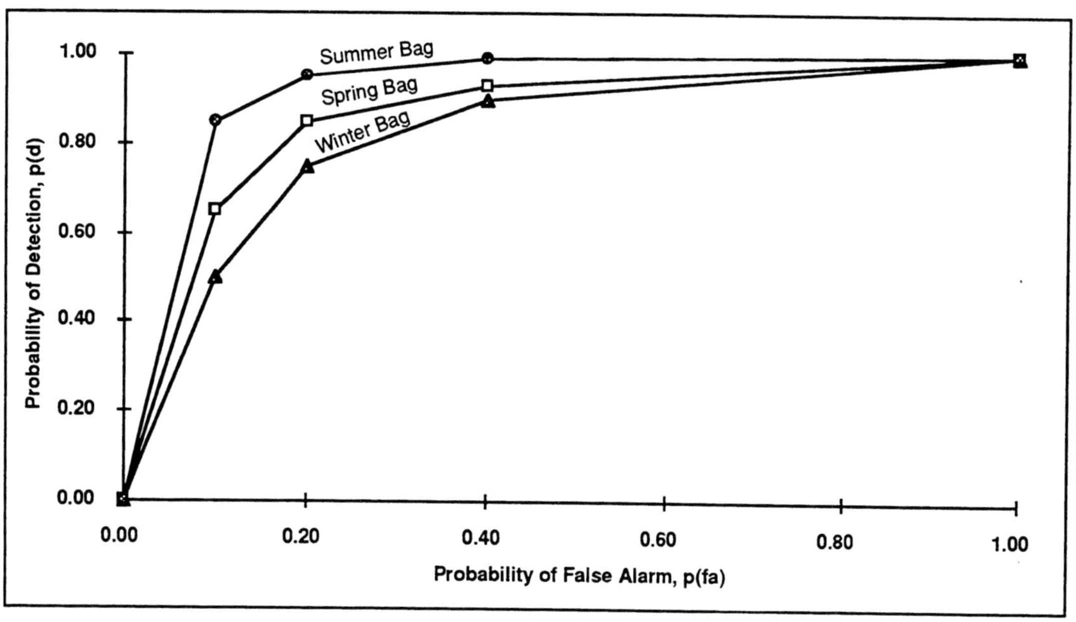

FIGURE A1. Seasonal Variation in TNA Detectivity Versus False Alarm Rate

ANNEX II

VALIDATION OF SIMULANTS

Annex II contains a detailed statistical analysis procedure for validation simulants.

<u>1. Test Procedures.</u> Explosives will be placed in bags representative of the total set of population of bags being tested. The bags are then inspected by the equipment, and the resultant values of each of the characteristics are recorded. Then, using the same set of bags but replacing the explosives in the same location with explosive simulant, repeat the test. Note: if possible, the selection of the location should be such as to provide high signal-to-noise ratio in order to facilitate the explosive-simulant comparison. All bags should be sampled at least ten times with the explosive, with the corresponding simulant and with neither. By comparing the average and the variances of the reading of the characteristics with explosives in the bag, with the average and the variances of the readings of the characteristics with the simulated explosive in the same bag, the validity of each simulant can be determined.

For example, in order to test simulants of ten explosive types identified by FAA, fifteen bags, covering the population range of the characteristics being measured, are recommended for use. Five empty bags (containing neither explosive nor simulant, but only the normal contents of the bag) would be randomly placed among the ten bags containing either explosives or simulants on every one of the major runs that are used to obtain data on the explosives and the simulants. One purpose of using the same empty bags on each run is to collect data to evaluate if the equipment has "memory," in that if higher values of characteristics are being measured on a given run (as a result of ten bags containing explosives or simulants) then the equipment might be reading higher on all bags, including the empties. A second purpose is to collect large quantities of data of a selected number of bags to estimate the variability of the measurements from the selected bag set. This should be used to reconfirm the bag selection process for validation testing and to help guide the test team in the bag selection process for testing at an airport terminal. The test team must monitor all aspects of the validation testing, record data for each bag processed, verify the sequencing of bags through the detection equipment, place all explosives and simulants in the appropriate bags, and observe the verification tests of each of the samples of explosives used for comparison against the simulants.

<u>2. Test Results</u>. The detailed data collected on each bag will be maintained by the FAA. The mean value and standard deviation of the measurements of the characteristics for empty bags, the bags with explosive and the same bags with the simulated explosive should be recorded and compared. For each

characteristic, its mean value for the bags with an explosive is compared to its mean value for the bags with the simulated explosive using conventional statistical analyses, such as F-tests and t-tests. A priori, the test team should select the critical region for both the F and t tests (i.e., reject the hypothesis that the simulant is representative of the explosive with respect to the characteristics being considered). This region is usually selected to be at the α=5 percent level.

Based on this data, simulants are accepted or rejected using the "t test." The rejected simulants may be reworked (if possible) and rerun through the equipment. This process may be continued in an attempt to validate as many simulants as possible. For overall confirmation, analysis of variance can be performed to determine if differences of the mean value of the characteristic of the bags with real explosives and the mean value of the characteristic of the bags with simulated explosives were zero (using the F-test, with the associated test of homogeneity). The detailed statistical approach that can be used is described in any statistical analysis book.

The validated simulants should be suitably marked and immediately placed in the custody of the test team who will then deliver them to the test site at the appropriate time. If the simulant characteristics can change over time, the simulants should be revalidated after appropriate time periods.

The decision to be made is whether or not a given simulant gives a similar enough response to that given by the real explosive as far as equipment operation is concerned, namely detect or no detect. The following testing is recommended for the validation process.

Start with a representative group of J+Q bags where J equals the number of different explosive types being simulated and Q represents a small number of additional bags. For the j-th explosive type, there are m(j) simulants which need to be validated. Here the m(j) depends on the test design. These J+Q bags are run through N times for each of the three conditions, (1) with the explosive, (2) with the matching simulants, and (3) with neither (to check on additivity of the measurements and to ensure that the bag characteristics have not changed over the testing period). The explosive and its matching simulant are also placed in the same bag in the same position, that position being where the measurements of the characteristics are considered to be optimal for measurement purposes.

The set of characteristic measurements of the bag are the determining operational variables used in the test. Thus the primary data can be considered as three matrices:

(1) a{i,k}
(2) a{i,E(j),k}, and
(3) a{i,S(j,m),k},

where (1) is the observed value of the i-th characteristic in the k-th bag, (2) is the observed value of the i-th characteristic with explosive type E(j) in the k-th bag, and (3) is the observed value of the i-th characteristic with the explosive replaced by the m-th simulant of the j-th type explosive, S(j,m), in the k-th bag. Here,

$$i=1,...,I \qquad j=1,...,J$$
$$k=1,...,K=J \qquad m=1,...,m(j)$$

The following process is recommended:

(i) Run K+Q bags through N=10 times without explosives or simulants. This data is needed to ensure that the bags being used are representative of the bag population they are thought to be representing and to check for additivity of the measurements;

(ii) Place the J explosives in the K bags, selecting the assignment at random, all in the same location, that being where the a{} readings are expected to be least variable. If appropriate, use the flip/twist move procedure, described above, to sample the a{} at different relative locations in the equipment. Run the K+Q bags through N=10 times;

(iii) Replace each of the J explosives with a simulant of that explosive, S(j,m), and repeat (ii), continuing this until all simulants have been tested in bags which contained the appropriate explosive;

(iv) At the beginning of each new day of testing and at the end of the testing, repeat (i).

3. Analyses for the Validation of Simulants

Let

$$A\{\}=\Sigma a\{\}/10$$

be the mean value of the respective reading, taken over the 10 observations on the k-th bag, and

$$v()=\Sigma[a\{\}-A\{\}]^2/9$$

be the sample variance.

In order to validate a simulant, we are testing the hypothesis that the population mean as estimated by A{i,E(j),k} is equal to the population mean as estimated by A{i,S(j,m),k}, given the populations are normally distributed with equal variances. Three statistical tests can be used to determine if these samples came from the same distribution, so as to justify the validation of the m-th simulant of the j-th type explosive.

a. The Significance of the Difference Between Two Sample Means.

In this approach only one bag is used for the testing of the explosive and the simulant of that explosive. That bag (say the k-th bag) should be randomly assigned for a given simulant and explosive pairing.

For a simulant of the j-th type explosive, one now tests that A{i,S(j,m),k}=A{i,E(j),k}, for all m=1,...,m(j) and for all characteristics i=1,...,I. If, for each of the I characteristics, the two sample means are not significantly different at the 1-α level of significance using the two-sided alternative, we will declare the m-th simulant to be validated, at the $1-(1-\alpha)^I$ level of significance.

To make such a statistical test, one must first determine if the two observable random variables, for each of the I characteristics, have equal variances or not. To test this one uses the Snedecor F statistic computing

$$F(9,9)=V(i,j,m)/v(i,j,m),$$

where by convention V() represents the larger of the two variances, and v() the smaller, and (9,9) are the respective degrees of freedom.

If the F-value proves to be not significantly different than 1, using once again the 1-α level of significance, one can pool the two variances into a single variance

$$v(i,j,m)=[9v(i,E(j))+9v(i,S(j,m))/18$$

Then the student t-test statistic is

$$t(18)=[A\{i,E(j),k\}-A\{i,S(j,m),k\}]/[v(i,j,m)*2/10]^{.5}$$

If $|t(18)|<t(18|1-\alpha/2)$, for each of the I characteristics, the simulant will be declared to be validated at the $1-(1-\alpha)^{I}$.

If, however, the F test of the two variances shows them to be significantly different, then the simulant does not exhibit the same properties as the explosive and should be rejected. In this validating approach each of the m simulants are considered independently even though some of them are simulating the same type of explosive.

b. <u>The Paired Difference Approach.</u>

Since the values of a{} come from the same bag with the explosive and its matching simulant in the same position in the bag, it is natural to consider the paired difference approach to test the correspondence between a simulant and the explosive.

In this approach, one uses the differences

$$d\{i,j,m\}=a\{i,E(j),k\}-a\{i,S(j,m),k\},$$

where the pairing over the 10 observations are randomly assigned. The population of such differences has an expected value, E(d{i,j,m})=0 if indeed the m-th simulant faithfully represented the j-th explosive for the i-th characteristic. An appropriate null hypothesis to test would be

H(0): E(d{i,j,m})=0,

against the two-sided alternative,

H(1): E(d{i,j,m})$\neq$ 0.

Once again the m-th simulant could be accepted as being validated if the null hypothesis is not discarded at the α level of significance for all of the I characteristics. The test statistic to use in making this test is:

$t(9)=D(i,j,m)/[v(d(i,j,m]^{.5}$

where $D(i,j,m)=\Sigma d(i,j,m)/10$, taken over the 10 samples, and

$v(i,j)=\Sigma[d(i,j,m)-D(i,j,m)]^2/9$.

Note that in using this test we have lost 9 degrees of freedom and as a result this test would be preferred only when there is a relatively high correlation between a{i,E(j),k} and a{i,S(j,m),k}. Here again, each simulant would be subject to its own validating decision.

c. <u>The Analysis of Variance Approach.</u>

It should be noted that d(i,j,m) as defined above, can be considered in an Analysis of Variance for each explosive type j. The Analysis of Variance approach provides a statistical test of the composite null hypothesis:

H(0): E[D(i,j,m)]=C, for all i and m

against the alternative,

H(1): E[D(i,j,m)]$\neq$C, for some i or m.

The Analysis of Variance table would simply be

<u>Analysis of Variance</u>

Source of Variation	Sum of Squares	Degrees of Freedom	Mean Squares
Among simulants	SS (Simulants)	m(j)-1	S_M
Within simulant (Experimental Error)	SSE	m(j)*(I-1)	S_P
Total	SST	I*m(j)-1	

Here

$$S_M=I*\Sigma(D(j,m)-D(j)^2/(m(j)-1)$$

$$S_p=\Sigma(\Sigma(D(i,j,m)-D(j,m))^2/m(j)*(I-1)$$

and

$$D(j,m)=\Sigma D(i,j,m)/I, \quad D(j)=\Sigma\Sigma D(i,j,m)/m(j)*I$$

We can then test the above hypothesis by selecting the level of significance α and assuming homogeneous variance, establish the critical region (reject H(0)) as

$$F>F_{1-\alpha}(m(j)-1, m(j)*(I-1))$$

where

$$F=S_M/S_p$$

This test can then be applied for every one of the J explosives.
Tests such as this are available on most computer's statistical packages.

If the null hypothesis is accepted, then one needs to test the hypothesis that the constant C=0. One approach to this test is the use of the t-test.

In using this test if the composite null hypothesis is not discarded all simulants of the j-th type explosive would be declared to be validated. However, if the null hypothesis is discarded, none of the simulants would be validated. There is an approach that allows one to also examine a wide variety of possible differences, including those suggested by the data itself. This test is based on the range of the m(j) observed means (we are looking at the difference between the smallest and the largest observed values). The reader is referred to Introduction to Statistical Analysis, Dixon & Massey, Second Edition, pages 152 to 155.

d. Measurement Additivity.

As was mentioned above only one bag is used to validate the simulants of a given explosive. This is acceptable if the measurement system is additive, that is, if the measurement of the characteristic of the explosive plus the measurement of the characteristic of the bag statistically equals the measurement of the explosive in the bag. If this is not the case, then the validation of the simulants will require testing over a representative set of bags, for each explosive threat. Hence one must first test for additivity of the measurements. This is done by taking measurements of the explosive in the absence of any background which would interfere with the measurement of the characteristics of the explosive. These measurements are repeated 10 times (as in the above). One then compares the sum of average reading of the explosive and the average reading of the bag (which is associated with the testing of that explosive, but not containing the explosive) with the average reading of the bag with the explosive. The above Analysis of Variance approach is suggested for testing for additivity of the measurements.

4. A Recommended Approach for Validating Simulants

Repeated use of the Analysis of Variance approach appears to be appropriate. If in its initial use, none of the simulants are validated, individual t-tests should be run and the simulant which differed most significantly from its null hypothesis would be declared to be invalidated and dropped from the group of simulants being tested. The Analysis of Variance approach would then be repeated using the reduced group. This process would be repeated until a final reduced group of simulants would be declared to be validated. This is then repeated for all of the J explosives.

ANNEX III

MODEL TESTING SCENARIO

The following testing scenario is a model that could be used for the actual testing. For explanatory purpose only, it will be assumed that 20 threats are available to conduct the test. If passenger bags cannot be used, then it is suggested that for each of the bag populations, six groups of twenty bags be selected as representative of that population. If it is not possible to "reproduce" the distribution because an adequate number of bags is not available, a statistical weighting process could be used to "match" the distribution.

After having selected the six groups of twenty bags, randomly order the bags in each group with the first group, called A1, A2, ..., A20, the second group called B1, B2, ..., B20, etc. To accomplish this, one can use a random number generator or a table. Next, the first 10 B bags are randomly intermingled with the first 10 A bags and the second 10 B bags with the second 10 A bags. Twenty threats are then randomly assigned to the 20 A bags. Each newly defined group of ten A and ten B bags are then processed through the equipment. Each time the group of twenty bags is processed, the threat location in the bag is changed by flipping and/or twisting the bag. Figure A2 shows the 15 possible locations in the bag and 9 locations that should be used since they represent all 15 locations.

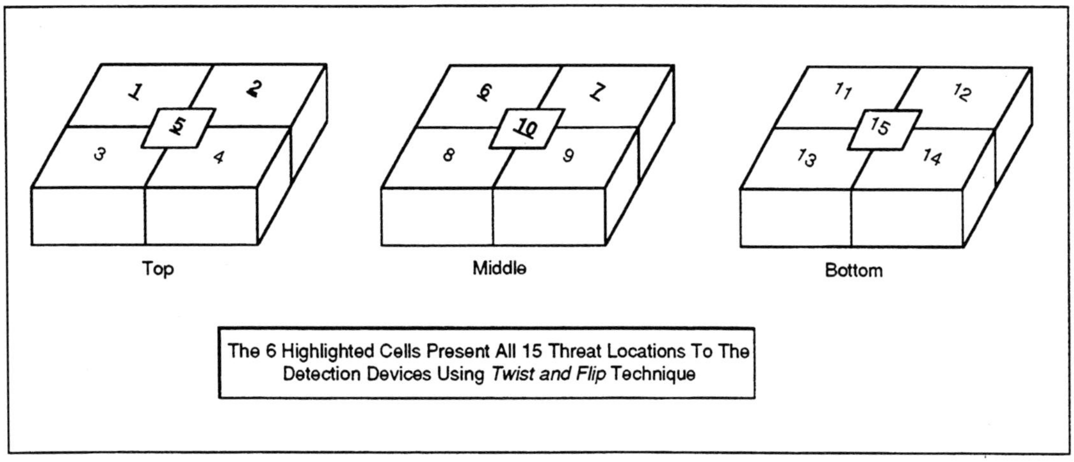

FIGURE A2. Threat Placement Locations

The threats are then transferred to the B bags in the group and again processed through the equipment. In this manner, one is able to obtain measurements on A bags with threats and B bags without and on B bags with threats and A bags without, obtaining 9 observations on each bag. Then testing proceeds with C & D bags (using the same process - replace A with C and B with D) obtaining 9 observations on each combination. Finally testing is done with E & F bags (using the same process--replacing C with E and D with F) obtaining 9 observations on each combination. The total number of observations on empty bags is 1080 while the number of observations on bags with threats is 1080, with each threat being tested in 6 different bags, and in 9 locations in each bag.

Before running the six groups of twenty bags through the equipment, two control bags should be processed 10 times, one bag without a threat and the second bag contained one of the threats. This set of two bags should be processed in a similar manner at various times during the tests and also at the end of each day. Comparing these measurements assure the test team that the equipment readings are not varying over the duration of the tests.

APPENDIX B

GLOSSARY OF TERMS

Bulk Detection	Sensing some physical or chemical property of an object under investigation. Usually used to differentiate from vapor detection which collects and analyzes volatized molecules.
Certification Test	A pass/fail test conducted by the FAA to determine if an EDS meets the EDS Performance Standard. The EDS must be a production version. If an EDS passes the certification test, it is qualified from an operational point of view to be deployed. It does not necessarily mean it is certified for deployment.
EDD	Explosive detection device. This is an instrument which incorporates a single detection method to exploit one or more common physical property of explosive materials for detection purposes.
EDS	Explosive detection system. This is a self-contained unit composed of one or more devices (EDDs) integrated into a system. It includes automatic detection, all necessary computer hardware and software, and baggage movement, positioning equipment, calibration methods, and maintenance procedures.
False alarm	An indication that explosive material is present when in fact there is no explosive material; also known as a false positive.
FFRDC	Federally Funded Research and Development Corporation. A not-for-profit private corporation which is chartered to only perform specialized work for federal agencies; prohibited from competing with private industry.
Full Level Threat	The full level (100 percent) threat is the minimum amount and configuration of a specific explosive material which would reasonably be expected to cause catastrophic damage to commercial aircraft in service.

Reference Instrument	Refers to an accepted or validated instrument that can be used to establish another instrument to give the correct reading.
Interferent	A substance that causes a response in a particular detection instrument that is similar enough to a real explosive that its response cannot be discriminated from a real explosive.
p(d)	An estimate of P_D derived from analysis of test data.. The formula is given by: p(d) = (# of detections)/(total number of possible detections)
p(fa)	An estimate of P_{FA} derived from analysis of test data.. The formula is given by: p(fa) = (# of false alarms)/((total # of possible false alarms)
Parametric Testing	Testing conducted to quantitatively measure critical performance parameters as a function of key variables.
P_D	True probability of detection. It is a measure of Sensitivity.
P_{FA}	False alarm rate; i.e. probability of sounding an alarm when no threat is present. Also known as false positives.
Qualification Test	An test conducted on an EDD or EDS to verify the performance claims of the manufacturer. Results can be in the form of parametric performance data.
R	Average time (in seconds) required to process a bag by an EDD or an EDS. This is affected by the instrumental method, the time required for the processing of the detection algorithms, and the bag transport system.
r	An estimate of R derived from analysis of test data. The formula is given by: r = (time required to process all bags)/(# of bags processed)
Sensitivity	Probability of detection of a threat, if a threat is present; i.e. a true positive. Same as P_D.
Specificity	Probability of not triggering an alarm if no threat is present; i.e. true negative. Calculated as $(1-P_{FA})$.

T&E	Test and Evaluation. This is the structured gathering of test data in accordance with a pre-planned strategy in response to specific items of interest. The test phase is followed by an analysis phase, which makes extensive use of statistical approaches to determine the values of measured parameters and the significance of the data.
Terrorism	Premeditated, politically motivated violence perpetrated against noncombatant targets by subnational groups or clandestine state agents, usually intended to influence an audience.
Verification Test	Same as Qualification Test.
Voxel	Volume element. Usually used to describe the element which a detection method samples and reports results. The number of voxels is a measure of resolution.
Z	Atomic Number. Which is equal to the number of protons in the nucleus. Isotopes of atoms have the same Z value, but different number of neutrons.

APPENDIX C

BIOGRAPHICAL SKETCHES OF COMMITTEE MEMBERS

JOHN D. BALDESCHWIELER received a B.Ch.E. degree from Cornell University and a Ph.D. degree in physical chemistry from the University of California at Berkeley. He was deputy director of the Office of Science and Technology, Executive Office of the President, and also served on the faculties of Stanford University and Harvard University. He was chairman of the Division of Chemistry and Chemical Engineering at the California Institute of Technology, where currently he is a professor of chemistry. Dr. Baldeschwieler is a member of the National Academy of Sciences. His research interests are molecular structure and spectroscopy including nuclear magnetic resonance, nuclear spectroscopy, mass spectroscopy, x-ray diffraction, and application of these methods to biological systems.

JONATHAN W. AMY received a B.A. in Chemistry from Ohio Wesleyan University and a PhD. in Molecular spectros, in 1955 from Purdue University. He was a research associate chemist, associate professor and director of instrumentation at Purdue. He is a emeritus professor of chemistry at Purdue University. His research interests include chemical instrumentation, and areas of molecular and mass spectroscopy, gas chromatography, and surface chemistry.

ALFRED BLUMSTEIN received a B. Eng. in Physics and a Ph.D. degree in operations research from Cornell University. After employment at the Cornell Aeronautical Laboratory and the Institute for Defense Analyses, he joined the faculty of Carnegie Mellon University where he currently is the J. Erik Jonsson Professor of Urban Systems and Operations Research and Dean of the School of Urban and Public Affairs. His research interests include quantitative and policy research in law enforcement and criminal justice; urban transportation; and family planning.

ARTHUR FRIES received a MA in Mathematics and a PhD. in statistics from the University of Wisconsin. He was a research and teaching assistant at Wisconsin University. He now works with the Institute for Defense Analyses (IDA) where he is a staff member and project leader. He also is a part-time lecturer for George Mason University and the University of Southern California. His research interests include areas of test planning and evaluation methodology of major weapons systems as well as in reliability, statistics, and applied mathematics.

STANLEY S. HANNA received an A.B. degree and an honorary D.Sc. from Denison University and a Ph.D. degree in physics from Johns Hopkins University. After employment at Johns Hopkins University and Argonne National Laboratory he became a professor of physics at Stanford University. He was chairman of the Division of Nuclear Physics of the American Physical Society (APS) and served on the

Executive Committee of the APS. His research interests include nuclear physics and structure; giant resonances; polarizations of nuclear radiations; lifetimes of nuclear states; resonance absorption; Mossbauer effect; nuclear moments; analog states; hyperfine interactions; magnetism; and electron scattering.

WILFRED A. (BILL) JACKSON received a B.S. in business administration from West Virginia University and a M.S. in management from George Washington University in 1954, and attended Industrial College of the Armed Forces at Ft. McNair, VA in 1974. He was with the U.S. Army for 26 years, before being employed by the BDM Corporation, Mitre Corporation, BWI Airport, before coming Director of Security and Facilitation, with the Airport Operators Council International until 1992, and has recently taken a position with the University of North Dakota as Assistant Professor for the Center for Aerospace Sciences. His research interests include airport operations, airport security, and computer-controlled access systems.

RICHARD H. JUDY, received a BA degree in airport management at the University of Miami in 1953. He was employed as an Office Manager for Merrill Lynch, worked as an auditor for Florida State, was a senior auditor with Morgan, Altemus and Barrs, was a comptroller for Dade Co. Port Authority, Biggs Shipping Corporation, Royal Castle Systems, State of Florida Road Department, was Deputy Director Dade Council Aviation Department, and before becoming President of Richard H. Judy and Associates, he was Director of the Dade County Aviation Department. He was chairman of the Airport Operator's Council International, and is a member of several associations. His research is in the development of domestic and international aviation regulations, policy and law pertaining to airports, airlines, the environment and the consumer.

BRUCE R. KOWALSKI received a B.A. degree in analytical chemistry, from Millikin University, and a PhD. from the University of Washington, Seattle. He was employed with the Shell Development Company as a chemist, was associate professor at Colorado State University, and is presently employed with the University of Washington, Seattle as associate professor. His research interests include areas of chemometrics, application of pattern recognition and other multivariant analysis methods to chemical data.

HAROLD MCNAIR received a degree B.S. from the University of Arizona in analytical chemistry, along with his M.S. and PhD. He was employed with Esso Research & Engineering as a research chemist, as a technical director with the European Division of F&M Science Corporation, Amsterdam; was a general manager and became director of international operations Varian Aerograph, Switzerland, and marketing director with the California branch. He is currently employed as a professor of analytical chemistry at Virginia Polytech Institute and State University. His research interests are extensive experience in quantitative analysis of ionization detectors and temperature programming, trace gas analysis by ionization detectors, theory of chromatography.

DAVID MILLIGAN received a B.A. degree from Princeton University and M.S. and Ph.D. degrees in chemistry from the University of Illinois at Urbana. After employment with 3M Company as well as Litton and Xonics Medical Systems he joined Abbott Laboratories where he is presently vice president of Diagnostic Products R&D. His research interests include photographic science immunodiagnostic systems and immunoassay techniques.

JOHN R. "Dick" ORR, attended the Georgia Institute of Technology, worked with Delta Airlines, in the technical operations division, on aircraft structure, was engineering director, and is presently manager of operations service administration. His research interests are operational security matters, both domestic and international for Delta Air Lines. He is involved with security equipment selection, procurement, installation, and operational procedures to comply with FAA regulations, as well as basic understanding of how a security program must be integrated in an airport community.

NORMAN SLAGG received a B.S. degree in chemistry from Brooklyn College and a Ph.D. degree from the Polytechnic Institute of Brooklyn, majoring in physical chemistry with an emphasis on kinetics. He obtained a fellowship to the National Bureau of Standards where he performed photochemical studies. After working for seven years in industry he obtained a position as a research chemist with the U.S. Army at Picatinny Arsenal. He is currently the chief of the Explosives and Warheads Branch. In addition, he has taught undergraduate physical chemistry part-time at Fairleigh Dickenson and Rutgers University for several years. His research interests include reaction kinetics, explosion phenomena, free radical reactions, thermal stability, and shock phenomena.

MICHAEL STORY received a B.S. degree in chemistry from the University of California, Berkeley. After graduation he worked at Litton Industries in material science of microwave tubes and at Electronic Associates in the design and manufacture of residual gas analyzers. He was a cofounder of Finnigan Corporation. He is presently employed as Senior Vice President, and Technical Director of Northwest Instrument Systems, Inc. His interests have been the research and development of mass spectrometers.

HARVEY E. WEGNER received a B.S. degree from the University of Puget Sound and M.S. and Ph.D. degrees in physics from the University of Washington. After employment at Brookhaven National Laboratory and Los Alamos National Laboratory, he returned to Brookhaven as a Senior Scientist to build the three-stage Van de Graaff accelerator facility for low-energy heavy-ion physics. He then helped design and develop the relativistic heavy-ion capability of the Brookhaven Alternating Gradient Synchrotron. His research interests include low-energy, heavy-ion accelerator design and construction, fast-neutron physics, low-energy reaction physics, and currently, relativistic heavy-ion physics. He has recently retired from Brookhaven.